6th International Conference on Mechanical and Aeronautical Engineering (ICMAE 2020)

IOP Conference Series: Materials Science and Engineering
Volume 1102

Chengdu, China
17 - 20 December 2020

Editors:

Dan Zhang
Ian McAndrew

ISBN: 978-1-7138-3396-3
ISSN: 1757-8981

Printed from e-media with permission by:

Curran Associates, Inc.
57 Morehouse Lane
Red Hook, NY 12571

proceedings.com

Some format issues inherent in the e-media version may also appear in this print version.

Institute of Physics
Dirac House, Temple Back
Bristol BS1 6BE UK

Phone: 44 1 17 929 7481
Fax: 44 1 17 920 0979

techtracking@iop.org

Additional copies of this publication are available from:

Curran Associates, Inc.
57 Morehouse Lane
Red Hook, NY 12571 USA
Phone: 845-758-0400
Fax: 845-758-2633
Email: curran@proceedings.com
Web: www.proceedings.com

TABLE OF CONTENTS

AERONAUTICAL ENGINEERING

MECHANICAL ENGINEERING

IOP Conf. Series: Materials Science and Engineering 1102 (2021) 011001 doi:10.1088/1757-899X/1102/1/011001

PREFACE

The entire world is still under the influence of the virulent pandemic COVID-19. Unfortunately, each of us is affected one way or another. Our conference, the 6th International Conference on Mechanical and Aeronautical Engineering (ICMAE 2020) was not an exception. In December we held our conference online and although the same it was a great success and we, as other conferences, continued in our sharing research success. This change to virtual was achieved by using Tencent VooV software.

Each keynote speech lasted 40 minutes, invited speech 30 minutes and authors presentation 15 minutes. Each presentation was included with questions and answers. that offered a chat function though the software used to increase participation. The success and prosperity of the conference is reflected high level of the papers received. ICMAE 2020 became an effective communication platform for all the participants over the world and unlike some that claim international reach this conference was truly international.

ICMAE 2020 was organized by Hong Kong Society of Mechanical Engineers. This conference aims to provide a platform for researchers and engineers to share their ideas, recent developments, and successful practices in Mechanical and Aeronautical Engineering. More than 40 participants attended the conference, they were from USA, UK, Italy, Russia, Colombia, Canada, India, Malaysia, Japan, China and more. Three renowned speakers given speeches about their latest research and reports. They are: Prof. Dan Zhang from York University, Canada; Prof. Ian McAndrew from Capitol Technology University, USA; Prof. Andrew Rae from University of the Highlands and Islands, UK. The conference also had 2 technical session and 1 poster sessions.

The conference proceeding is a compilation of the accepted papers and represent an interesting outcome of the conference. This book covers 2 chapters: 1. Aeronautical Engineering; 2. Mechanical Engineering.

We would like to acknowledge all of those who supported ICMAE 2020. Each individual and institutional help were very important for the success of this conference. Especially we would like to thank the committee chairs, committee members and reviewers, for their tremendous contribution in conference organization and peer review of the papers.

We sincerely hope that ICMAE 2020 will be a forum for excellent discussions that will put forward new ideas and promote collaborative research and support researchers as they take their work forward. We are sure that the proceedings will serve as an important research source of references and the knowledge, which will lead to not only scientific and engineering progress but also other new products and processes. Finally, we hope to meet you all in person at our next conference in 2021.

Editors: Dan Zhang, York University, Canada
Ian McAndrew, Capitol Technology University, USA

COMMITTEES LIST

Conference Co-Chairs

Dan Zhang, York University, Canada

Weijun Pan, Civil Aviation Flight University of China, China

Program Committee Chair

Yong Zhao, Nazarbayev University, Kazakhstan

Scientific Committee

Andrew Rae, University of the Highlands and Islands, UK

Erasmo Carrera, Politecnico di Torino, Italy

Giovanni Belingardi, Politecnico di Torino, Italy/Automotive Research and Sustainable Mobility Center (CARS Center), Italy

Y. Lawrence Yao, Columbia University, USA

Yu Zhang, Zhejiang Univerisity, China

Organizing Committee

Ian McAndrew, Capitol Technology University, USA (Co-Chair)

Kyle Jiang, University of Birmingham, UK (Co-Chair)

Zhongbin Wang, China University of Mining and Technology, China (Co-Chair)

Chenglong Li, Civil Aviation Flight University of China, China

Jianping Xiang, Changsha University of Science and Technology, China

Hui Zhao, Zhengzhou University of Aeronautics, China

Financial Chair

Weimin Huang, Nanyang Technological University, Singapore

International Technical Committee

Angelo Bonfitto, Politecnico di Torino, Italy

Chunming Li, China University of Petroleum (East China), China

Eltayeb Eljack, University of Khartoum, Sudan

Jian Jun Zhu, Sr. CAE in vehicle dynamics and simulation at General Motors, USA

Jiang Ding, Guangxi University

IOP Conf. Series: Materials Science and Engineering 1102 (2021) 011002 doi:10.1088/1757-899X/1102/1/011002

Peer review declaration

All papers published in this volume of *IOP Conference Series: Materials Science and Engineering* have been peer reviewed through processes administered by the Editors. Reviews were conducted by expert referees to the professional and scientific standards expected of a proceedings journal published by IOP Publishing.

- **Type of peer review: Double-blind**
- **Conference submission management system: CMT**
- **Number of submissions received: 43**
- **Number of submissions sent for review:** 39
- **Number of submissions accepted: 25**
- **Acceptance Rate (Number of Submissions Accepted / Number of Submissions Received X 100): 64%**
- **Average number of reviews per paper: 2**
- **Total number of reviewers involved: 14**
- **Any additional info on review process:**

All the papers will be peer reviewed by 2-3 experts. It takes 20-30 working days for the result coming out. If the paper needs revising, it should be resubmitted for peer review again.

1. Manuscripts must be written in English;

2. The manuscript should be written in accordance with the standard of template;

3. The topics of paper should be relevant to conference topics;

4. Plagiarism and Duplicate submission are prohibited;

5. Innovation and scientific value is a must.

- **Contact person for queries: lynlee@smehk.org**

Research on UAV Flight Test Technology Based on Embedded Commands

Wenqiang Yin[a, *], Yalong Wang[b] and Shenjun Qi[c]

Chinese Flight Test Establishment, Xi'an, 710089, China

[a]yin19890510@126.com; [b]yalong2015@163.com; [c]64073907@qq.com

Abstract. This paper presents the current situation that the pilot's manual operation method can't fully meet the requirements of UAV flight test. To solve the problem, a UAV flight test technology based on embedded command is proposed. By analysing the verification contents and assessment methods of UAV flight test, the typical actions in UAV flight test are extracted, the embedded command modules are designed, and the embedded command flight test method is formulated. The flight test results show that compared with the manual operation method, the embedded command flight test technology can alleviate pilot's operation burden, reduce the flight test risks and improve the flight test efficiency.

1. Introduction

Generally speaking, the UAV system is composed of UAV platform, ground control station, data link system and ground support system. Compared with the manned aerial vehicle flight test, the UAV flight test has both similarities and uniqueness. The UAV platform as the basis and key component of UAV system, evaluating its function and performance by flight test, is an effective means to verify whether the UAV system meets the expected requirements [1]. During the flight test of UAV system, pilot's manual operation relies on the ground control station and date link system. As the pilot is separated from the aircraft platform during the flight, he couldn't sense the flight status. In addition, because of the existence of link time delay, the traditional method of manned aircraft flight test has some limitations and deficiencies when applied to the UAV flight test [2].

Based on requirements of UAV platform flight test, this paper sorts out the UAV platform flight test verification project and establishes the cell library for the UAV typical flight test actions. By analysing the characteristics of UAV flight control system and the limitations of traditional manual stick-rudder control method, a flight control method based on typical flight test action commands is proposed. This method designs embedded commands for typical flight test actions, and proposes a verification scheme which is based on the UAV flight control system for typical flight test actions. The flight test results show that the method has good effect on the execution and completion of flight test actions when compared with the traditional manual stick-rudder control method.

2. Status of embedded commands flight test

From 2001 to 2018, the U.S. Department of Defense has issued 8 versions of unmanned systems integrated roadmap, which are used to guide the overall development of the U.S. military in the field of UAV system [3].During this period, Great progress has been made from the basic technology to the demonstration project of UAV. Among which the most famous ones are the HALE (high altitude long endurance) UAV represented by Global Hawk and the MALE (middle altitude long endurance) UAV represented by predator [4-5]. Compared with the rapid development of UAV technology, the relative research on flight test methods and assessment standards is slower; most of them still refer to the methods of manned aircraft, there are deficiencies and defects which can't meet the requirements of UAV flight test [6].

According to the current design ideas of UAV system, there are great differences in system configuration between UAV and manned aircraft. During the flight test of UAV system, pilot's manual operation relies on the ground control station and date link system, as the pilot is separated from the aircraft platform, he couldn't sense the flight status, and because of the existence of link time delay, the pilot's manual operation quality is not high, the effect of flight data is also bad. Some actions even need to be done by repeating the flight test many times, which will lead to low test efficiency and increase the test cost [7]. Meanwhile, due to the complexity of UAV system, its reliability is relatively fragile, so the flight safety depends on the design of flight test method. When the pilot's manual operation intervenes in the control, it will increase unnecessary risks. Therefore, it is of great significance to complete the execution and operation of typical flight test actions by embedded instructions in UAV platform flight test. The United States has carried out the embedded command research in many UAV models. Early example is the in-flight stability analysis techniques in X-29A vehicle based on the embedded command [8-9] .More recently, in the X-43A and NF-15B vehicle flight test, the automated injection of multisine excitations was used to simultaneously characterize the stability of multiple axes with postflight analysis [10-11], in the X-48B aircraft flight test, a set of inner loop control system based on dynamic inverse was developed, and multiple exciting actions were embedded [12].

3. Establish the cell library of typical flight test actions

Like manned aircraft, the flight test assessment scope of UAV platform can be also divided into flight performance test, flight quality test, and flight control system test, based on these three specialties, this paper sorts out the flight test verification items and typical flight test actions.

For the flight performance specialty, the flight test assessment items mainly include airspeed system calibration, take-off performance, climbing performance, practical ceiling, maximum level flight speed, minimum level flight speed, endurance performance, glide performance, manoeuvrability, hover performance, landing performance, mission performance, etc. The typical flight test action cells that can be decomposed are stable level flight, level flight acceleration, level flight deceleration, fast climb slow climb, fast glide, slow glide, stable turn left, stable turn right, stable hover, constant heading angel flight, constant track flight, dive acceleration, altitude remote adjustment control, speed remote adjustment control, etc.

For the flight quality specialty, the flight test assessment items mainly include longitudinal static stability, longitudinal dynamic stability, lateral-directional static stability, lateral-directional dynamic stability, transient characteristics, residual oscillation, artificial simulated landing, etc. The typical flight test action cells that can be decomposed are stable level flight, level flight acceleration, level flight deceleration, longitudinal pulse, longitudinal multiple pulse, longitudinal 3211, longitudinal sweep frequency, longitudinal step, lateral step, lateral sweep frequency, course step, coordinated sideslip, lateral multiple pulse , course multiple pulse , lateral-directional combined multiple pulse, etc.

For flight control system specialty, the flight test items mainly include basic function flight test, limit function flight test, altitude hold accuracy, speed hold accuracy, side deviation hold accuracy, pitch angle hold accuracy, roll angle hold accuracy, heading angle hold accuracy, landing control accuracy, etc. The typical flight test action cells that can be decomposed are altitude remote adjustment control, speed remote adjustment control, side deviation remote adjustment control , constant roll angle flight, constant pitch angle flight, constant heading angle flight, constant track flight , etc.

By sorting and summarizing the above-mentioned typical flight test action cells, a library of typical flight test action cells for UAV platform assessment was established. The cell library contains stable level flight, level flight acceleration, level flight deceleration, fast climb, slow climb, fast glide, slow glide, stable turn left, stable turn right, stable hover, constant heading angle flight, constant track flight, dive acceleration, longitudinal pulse, longitudinal multiple pulse, longitudinal 3211, longitudinal sweep frequency, longitudinal step, lateral step, lateral sweep frequency, course step, coordinated sideslip, lateral multiple pulse, course multiple pulse , lateral-directional combined multiple pulse, altitude remote adjustment control, speed remote adjustment control, side deviation remote adjustment control, constant roll angle flight, constant pitch angle flight, etc.

4. Design of embedded commands

4.1. Classification and control strategy design of UAV typical flight test actions

UAV uses embedded command method for the purpose of achieving the action excitation in a standard, normative and effective way, obtaining the high-quality test data, and ensuring the controllability of UAV flight state.

In the process of typical flight test action commands design, the action cells in the cell library firstly need to be classified according to the action types. In this paper, the typical flight test actions are divided into four types: level flight acceleration and deceleration commands, remote control commands, model excitation commands and attitude control commands.

4.1.1. Level flight acceleration and deceleration commands. The level flight acceleration and deceleration commands include level flight acceleration command and level flight deceleration command. For the level flight acceleration command, the control strategy is to switch the throttle to the open-loop control state, then push the throttle from the current position to the rated state quickly, the calibrated airspeed will increase to the specified value, finally turn into the speed closed-loop control, and then switch the throttle to closed-loop state after a set time. For the level flight deceleration command, the control strategy is to switch the throttle to the open-loop control state, then push the throttle from the current position back to the idle state quickly, the calibrated airspeed will decrease to the specified value, finally turn into the speed closed-loop control, and then switch the throttle to closed-loop state after a set time.

4.1.2. Remote control commands. The remote control commands include altitude remote adjustment control command, speed remote adjustment control command and side deviation remote adjustment control command. For the altitude remote adjustment control command, the control strategy is to record the altitude value at the current moment, then on the basis of the flight altitude limit conditions of UAV, add an increment as the expected value, and finally switch to the altitude remote adjustment control command. For the speed remote adjustment control command, the control strategy is to record the speed value at the current moment, then on the basis of the flight speed limit conditions of UAV, add an increment as the expected value, and finally switch to the speed remote adjustment control command. For the side deviation remote adjustment control command, the control strategy is to record the side deviation value at the current moment, then on the basis of the flight side deviation limit conditions of UAV, add an increment as the expected value, and finally switch to the side deviation remote adjustment control command.

4.1.3. Model excitation commands. The model excitation commands include longitudinal step command, lateral step command, course step command, longitudinal pulse command, longitudinal multiple pulse command, lateral multiple pulse command, course multiple pulse command, longitudinal 3211 command, longitudinal sweep frequency command, lateral sweep frequency command and lateral-directional combined multiple pulse command. For the longitudinal step command, the control strategy is to disconnect the normal overload circuit in the flight control system, and inject a rectangular wave signal. For the lateral step command, the control strategy is to disconnect the roll angle rate circuit in the flight control system, and inject a rectangular wave signal. For the course step command, the control strategy is to disconnect the sideslip angle circuit in the flight

IOP Conf. Series: Materials Science and Engineering 1102 (2021) 012001 doi:10.1088/1757-899X/1102/1/012001

control system, and inject a rectangular wave signal. For the longitudinal multiple pulse command, the control strategy is to disconnect the normal overload circuit and inject a longitudinal multiple pulse signal. For the lateral-directional combined multiple pulse command, the control strategy is to disconnect the sideslip angle circuit and the roll angle rate circuit, and inject a lateral-directional combined multiple pulse signal. The operation principle of other pulse commands is basically the same, which is not listed.

4.1.4. Attitude control commands. The attitude control commands include fast climb, slow climb, fast glide, slow glide, stable turn left, stable turn right,, stable hover, dive acceleration , stable level flight, coordinated sideslip, constant roll angle flight, constant pitch angle flight, constant heading angle flight, constant track flight, etc. The control strategy can be divided into four categories: altitude hold control, roll angle control, pitch angle control and heading angle control. For the altitude hold control, the control strategy is to record the altitude value at the current moment, then on the basis of the flight altitude limit conditions of UAV, add an increment as the expected altitude command value, switch to the altitude hold mode, after a set time, and finally switch back to normal status. For the roll angle control, the control strategy is to record the roll angle value at the current moment, then on the basis of the flight roll angle limit conditions of UAV, add an increment as the expected roll angle command value, switch to the roll angle hold mode, after a set time, and finally switch back to normal status. For the pitch angle control, the control strategy is to record the pitch angle value at the current moment, then on the basis of the flight pitch angle limit conditions of UAV, add an increment as the expected pitch angle command value, switch to the pitch angle hold mode, after a set time, and finally switch back to normal status. For the heading angle control, the control strategy is to record the heading angle value at the current moment, then on the basis of the flight heading angle limit conditions of UAV, add an increment as the expected heading angle command value, switch to the heading angle hold mode, after a set time, and finally switch back to normal status.

4.2. Implementation principle of embedded command

In order to simulate the manual operation by embedded command control method, it is necessary to add the flight test action command module before flight control system. Figure 1 shows the schematic diagram of embedded command implementation.

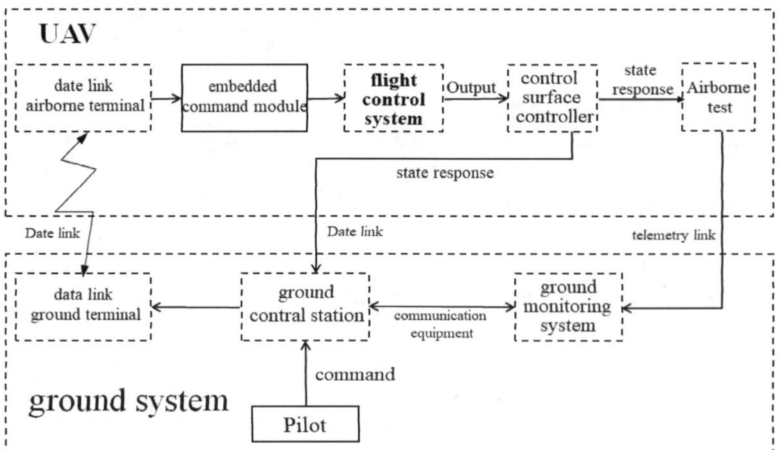

Figure 1. Schematic diagram of embedded command implementation

During the implementation, the pilot initiates the action command in the ground control station, and the command is transmitted to the date link airborne terminal through the data link, which triggers the embedded command module at front of the flight control system. After the embedded command is processed by the flight control system, the steering surface deflection command is output to the control

IOP Conf. Series: Materials Science and Engineering 1102 (2021) 012001 doi:10.1088/1757-899X/1102/1/012001

surface controller. On the one hand, the state response of the UAV is transmitted to the ground control station through the UAV's own data link which is used for pilot monitoring; On the other hand, the state response can also be transmitted to the ground monitoring system for technicians to observe through the additional telemetry link. The ground monitoring system and the ground control station can be connected through communication equipment.

According to the control characteristics of UAV system, the design of embedded command modules needs to be achieved from the three aspects: speed control design, attitude control design and trajectory control design. For the speed control design, it is based on the automatic throttle control loop. In the design of speed related commands (such as level flight acceleration and deceleration commands), the throttle is in the open-loop state first, and the desired speed is taken as the command value to control the engine throttle. When the speed reaches the specified value, it turns to the speed closed-loop control state. After a set time, switch back to the throttle closed-loop status. The speed control loop diagram is shown in Figure 2.

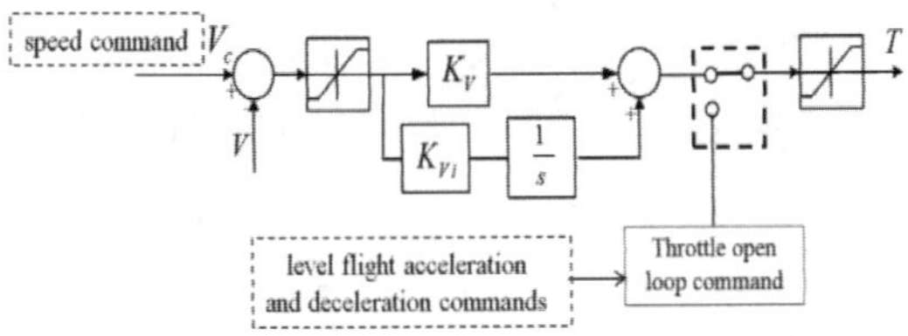

Figure 2 (a). Automatic throttle control loop

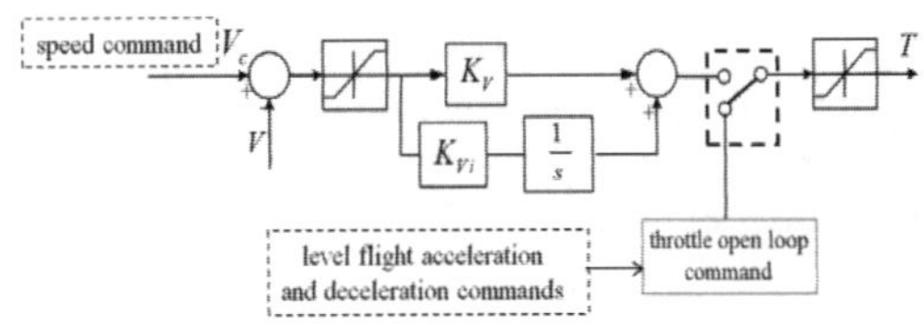

Figure 2 (b). Level flight acceleration and deceleration command control loop

For the attitude control design, it is divided into pitch hold control, roll angle hold control, and heading angle hold control. When designing the UAV attitude control command, the desired attitude is used as the command value, based on the current attitude, the UAV automatically calculates the attitude change and controls the UAV to respond to the input command. The pitch hold control is used as an inner loop of longitudinal control, which improves the damping ratio and longitudinal stability of UAV. When the throttle is constant, the pitch angle of UAV is held by adjusting the elevator. The pitch hold control loop is shown in Figure 3 (a). The roll angle hold control is the same as that of pitch hold control. The angular rate compensation is used as the inner loop controller to feedback the roll angel rate, so as to increase the damping and improve the control performance. The outer loop of roll angle hold control is angular position feedback. The roll angle of UAV is measured by vertical gyro sensor, forming a closed loop for series control. The roll angle hold control loop is shown in Figure 3 (b). The control law of heading angle hold control takes the lateral roll angle hold control loop as the inner loop; the heading angle hold control loop is shown in Figure 3 (c).

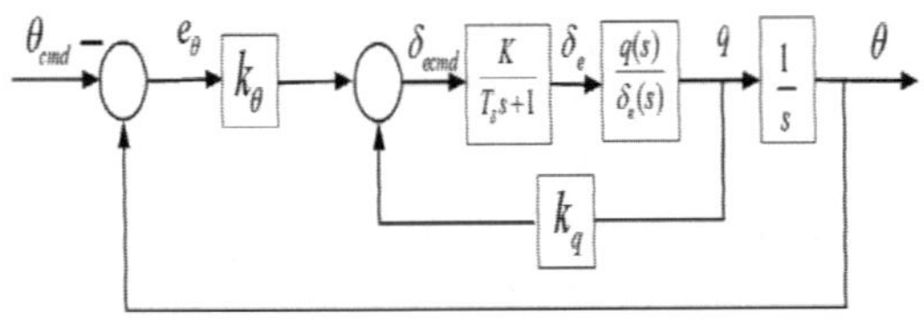

Figure 3 (a). Pitch hold control loop

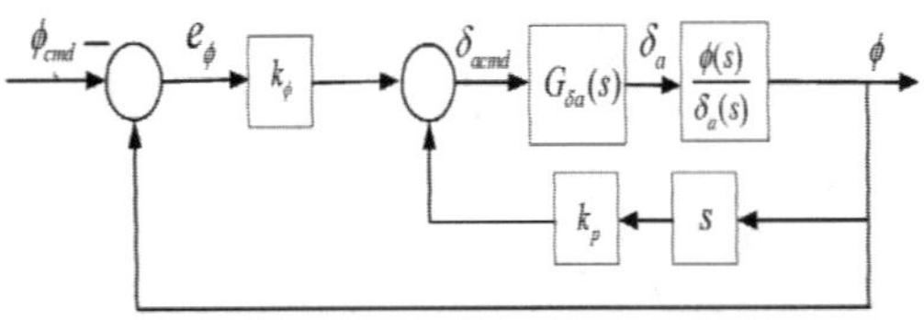

Figure 3(b). Roll angle hold control loop

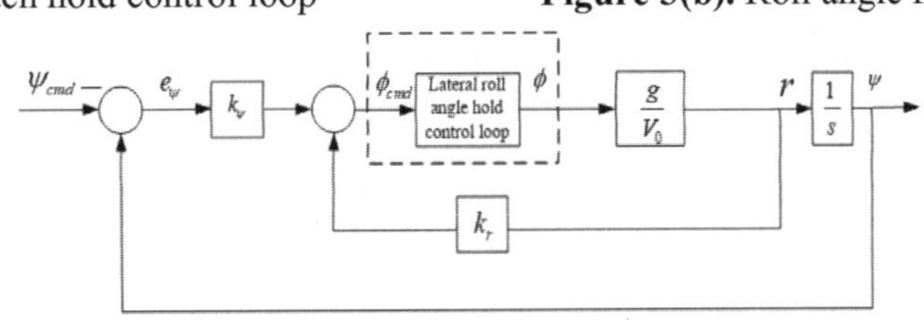

Figure 3 (c). Heading angle hold control loop

For the trajectory control, it mainly includes altitude hold control and lateral deviation control. When designing the command, the desired altitude or desired lateral deviation is taken as the command input, based on the current altitude or lateral deviation, the UAV automatically calculates the change of altitude or lateral deviation and controls the UAV to respond to the input command. As a part of longitudinal control, the altitude hold control loop of UAV takes the pitch angle hold control loop as inner loop, and designs with altitude change rate and altitude variation as feedback. The altitude hold control loop is shown in Figure 4 (a). The lateral deviation control loop uses the lateral roll angle hold control loop as the inner loop, as is shown in Figure 4(b).

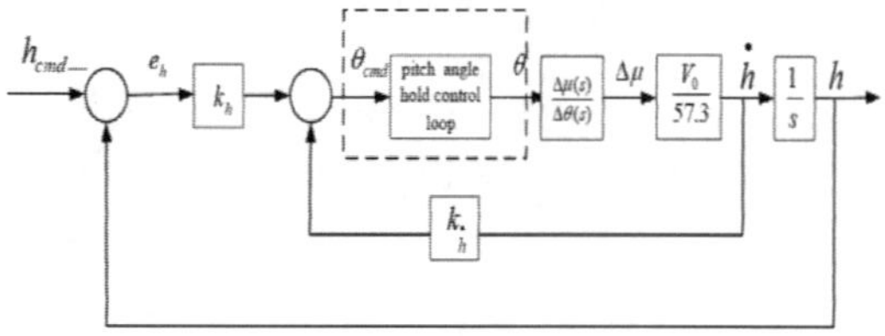

Figure 4(a). Altitude hold control loop

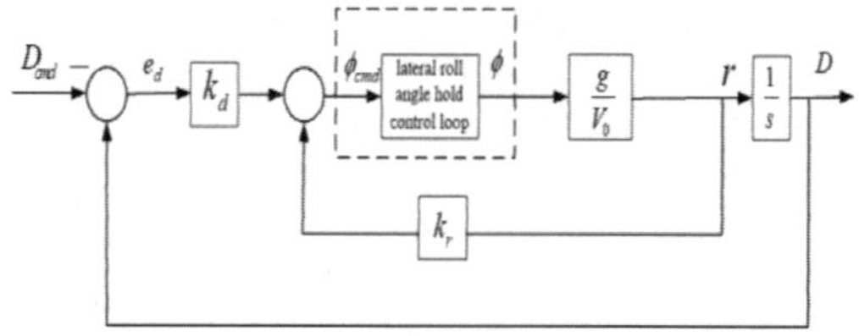

Figure 4(b). Lateral deviation control loop

5. Test verification of embedded commands

After the design of the embedded commands is completed, in order to ensure the safety of the flight test, it is necessary to carry out sufficient ground tests before the flight test. The verification contents of the ground tests mainly include checking the rationality of embedded commands design, checking the cross-linking matching between embedded commands and flight control system, preliminarily designing the scope of excitation signals, carrying out the function inspection and the fault simulation drill of embedded commands, etc.

In this paper, an UAV platform is selected for the embedded command flight test verification. In the verification, the flight test actions are carried out by manual operation method and embedded command method respectively. Due to the space limitations, four typical flight test results are given here, as shown in Figure 5 to fig Figure 8. The selected typical flight test actions include level flight acceleration, longitudinal multiple pulse, lateral-directional combined multiple pulse and stable hover.

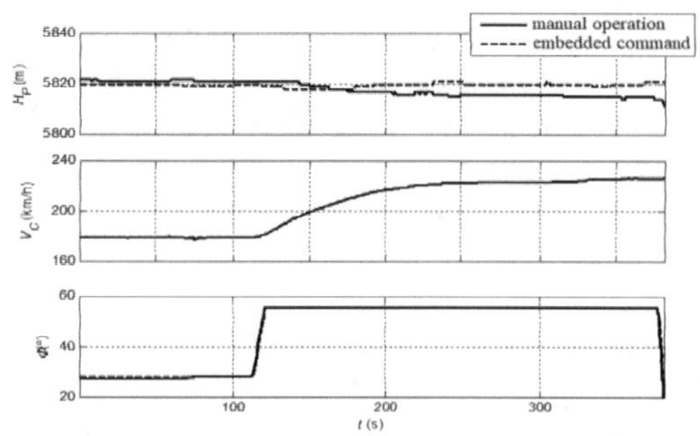

Figure 5. Level flight acceleration

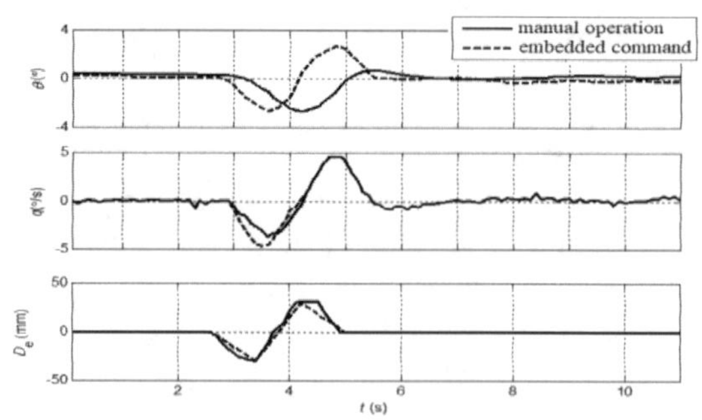

Figure 6. Longitudinal multiple pulse

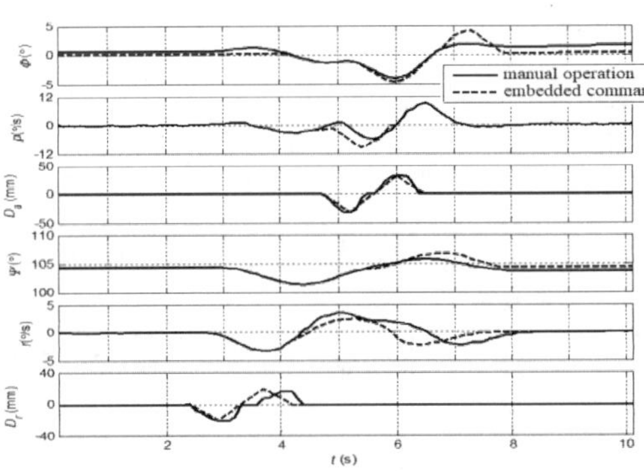

Figure 7. Lateral-directional combined multiple pulse

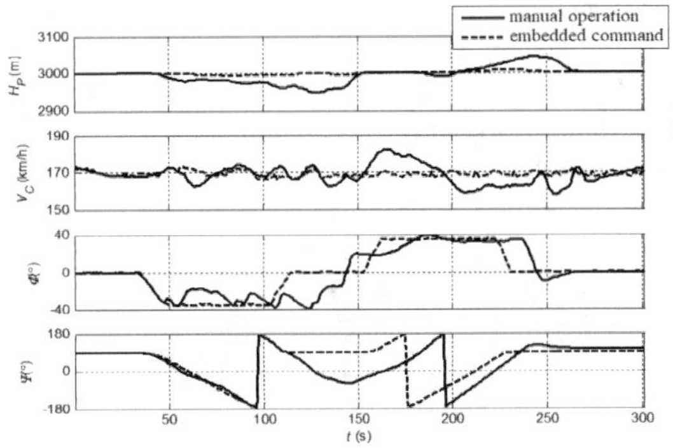

Figure 8. Stable hover

As can be seen from Figure 5 to Figure 8, for the level flight acceleration action, the control principle of manual operation method and embedded command method are the same which demands to push the throttle stick deflection to the rated state quickly, so the throttle stick deflection and the flight speed change trend of the two control methods are basically the same. However, it can also be seen that the embedded command method for the altitude control is obviously better than manual operation method; for the longitudinal multiple pulse action and lateral-directional combined multiple pulse action, the results show that the control effect of embedded command method on rudder deviation is significantly higher than that of manual operation control, the changing effect of attitude angle and attitude angle rate induced by embedded command method is significantly higher than that of manual operation method, and the data of flight test is more conducive to the flight quality evaluation; for the stable hover action, the embedded command method is also significantly better than the manual operation method in the altitude, speed and roll angle control, which is more conducive to the completion of UAV flight test tasks.

6. Conclusion

In view of the limitations of manual operation method, this paper proposes the embedded command technology based on typical flight test actions of UAV. By sorting out the UAV platform flight test items, a typical flight test action cell library is established, and the embedded command control modules and the flight test verification schemes are designed. The flight test results show that compared with the manual operation method, the embedded command method can alleviate pilot's operating burden, reduce the flight test risks and improve the flight test efficiency.

References

[1] Ang H S 2018 The concept and key technology of UAV system *Unmanned Systems Technology* **01** 66-71

[2] Qi S J, Wu X L and Sun J 2015 Current status and challenges of UAV airworthiness development *Aerodynamic Missile Journal* **07** 25-30

[3] Department of Defense 2018 *Unmanned Systems Integrated Roadmap 2017-2042* (Washington: Office of the U.S. Department of Defense)

[4] Reg A 2010 *Unmanned Aircraft Systems UAVs Design, Development and Deployment* (UK: WILEY) pp 303-316

[5] Richard K B, Stephen B H, Douglas M M and Eric S 2012 *Introduction to Unmanned Aircraft Systems* (UK: CRC Press) pp 6-16

[6] Chen C, Huang Y K and Zeng J 2012 Research on the standards of UAV system terminology *Aeronautic Standardization and Quality* **05** 43-46

[7] Wang D Y and Li Y L 2009 Key technology of high altitude and long endurance unmanned reconnaissance aircraft *Aerodynamic Missile Journal* **11** 31-35

[8] Bosworth J T and West J C 1986 Real-Time Open-Loop Frequency Response Analysis of Flight Test Data *AIAA* 1986-9738
[9] Gera J and Bosworth J T 1987 Dynamic stability and handling qualities tests on a highly augmented, statically unstable airplane *NASA* TM-88297
[10] Baumann E 2006 Tailored excitation for frequency response measurement applied to the X-43A flight vehicle *AIAA* 2006-638
[11] Morelli E A 2003 Multiple input design for real-Time parameter estimation in the frequency domain *13th IFAC Symposium on System Identification* REG-360
[12] Chirstopher D R 2008 In-flight stability analysis of the X-48B aircraft *AIAA* 2008-6571

The Design of Static Test Scheme for Composite Aileron Structure of Large Aircraft

Guanbiao Zhang[1,a *], Shuaicai Liu[1,b], Limeng Zhang[1,c] and Jiying Zhang[2,d]

[1] Composite Center, COMAC, Shanghai, China
[2] Requirement Engineering Development, COMAC, Shanghai, China

[a]zhangguanbiao@comac.cc; [b]liushuaicai@comac.cc; [c]zhanglimeng@comac.cc; [d]zhangjiying@comac.cc

***Abstract*:** According to the load conditions and structural features of composite aileron structure of large aircraft, the design techniques of full-scale static test scheme of aileron is studied, which can be used to examine the static strength and verify the airworthiness compliance of aileron structure. The overall test scheme design was carried out from the aspects of test load, test support, damage introduction and measurement scheme, and created a set of test design processes that meet the requirements of airworthiness. Loads screening was carried out by using strain coverage principle, the aerodynamic loads were transformed into test load by coordinate transformation. The wing was used as the aileron test support section, and a ground support device can also be designed to carry out aileron bench test. The introduction mode of aileron structure damage was defined, including the type, location and size. A reasonable layout of strain and displacement sensors was given for test measurement. Test results showed that the static test scheme in this paper can be used for the airworthiness compliance verification of static strength of aileron structure.
Keywords: Composite, aileron, static test, test scheme, airworthiness verification

1. Introduction

Composite materials have the advantages of high specific strength, high specific stiffness and good designability. They have become the basic materials of aerospace structures and are widely used in aircraft structures. The proportion of composite materials in aircraft structure has become one of the important indicators to measure the advanced nature of commercial aircraft. At present, composite materials are widely used in ailerons and other control surface structures in advanced large aircraft, which can reduce the structure weight on the basis of meeting the strength and stiffness. However, composite structures are sensitive to inherent defects and damage, and the failure modes are diverse and complex. At present, there is a lack of reliable analysis methods, which makes the reliable application of composite materials in the main structures of large aircraft face great challenges[1-2].

AC20-107B gives the building block test verification method[3] as an acceptable verification method for composite structure verification (Fig. 1). Full scale test is at the highest level of building block test. It can be used to verify the correctness of FEM and overall verify the structural strength. It plays an important role in the process of airworthiness verification.

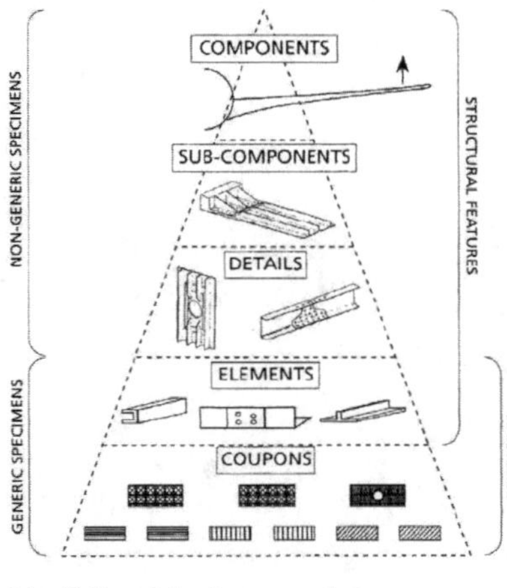

Figure 1. Classical building block test scheme presented in AC20-107B

In this paper, the full-scale static test scheme design of composite aileron structure of large aircraft is studied in detail. The test scheme design was carried out from the aspects of test load, test support, damage introduction and measurement scheme, and a set of test design process was summarized, which can be used as reference for the test of composite structure components.

2. Aileron Structure

Aileron is usually installed at the trailing edge of the wing near the wing tip, and is connected with the wing box through hinge joints and actuators. The main function of aileron is to control the rolling motion of the aircraft (Fig.2). Ailerons of large aircraft are generally wedge-shaped structures made of composite materials and aluminum alloy materials. Composite structures usually include skin, beam and ribs, joints are generally aluminum alloy structures.

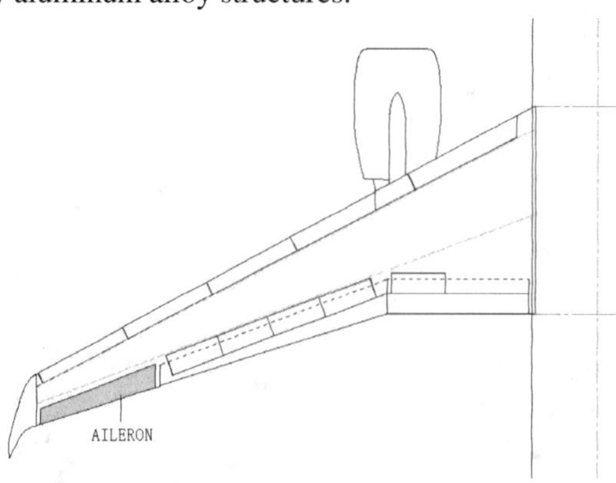

Figure 2. Aileron layout of large aircraft

The aileron is generally a fixed axis rotating structure. The hinge axis of the aileron is formed by the hinge joints. The aileron can be deflected up and down through the extension of the actuator. The serious loads to be considered for ailerons are divided into three cases: upward deflection, downward deflection and loads parallel to the hinge line. The load parallel to the hinge line is the load condition specified in CCAR25.393[4]. The load magnitude is small, and the stress level of the aileron is low, which mainly corresponds to the severe condition that the joint bears lateral load.

When the aircraft flies horizontally, the aileron is in the state of 0° and bears the aerodynamic lift. When the aircraft rolls, one aileron deflects upward and one aileron deflects downward. The lift of the aileron with downward deflection increases, and the lower skin bears larger upward aerodynamic load. The load of the upward deflection aileron becomes downward, and the upper skin bears the downward aerodynamic load (Fig.3). The downward aileron load increases on the basis of the original lift force, resulting in a larger load in the down deflection condition, and the most serious condition of the aileron is the down bias condition.

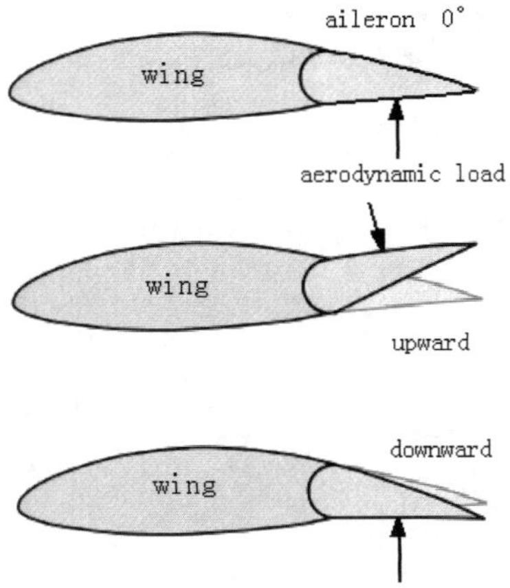

Figure 3. Aileron loading states

3. Test Support Scheme

In order to carry out the test scheme design, it is necessary to define the supporting state of the test. The support of the test should reflect the load and force transfer of the actual structure in flight. The aileron is connected with the wing beam through multiple joints. In flight, the aileron will be driven to coordinate deformation through the joint after the wing is deformed under load[5]. That is to say, even if the aileron is not loaded, the aileron skin will generate random deformation strain, which will affect the aileron strain distribution. The specific influence size and whether it can be ignored need to be determined by simulation analysis[6].

In order to analyze the influence of wing deformation on aileron, the finite element model of aileron internal force was used to calculate the aileron's strain. The load case with the maximum hinge moment of aileron was selected, and the matching load was applied on the wing, and the aileron was not loaded. The influence of wing deformation on aileron strain under this condition can be obtained through FEM. In general, the influence of wing deformation cannot be ignored, and the influence of wing deformation should be considered in aileron test support scheme.

There are two kinds of support schemes which can realize the coordinated deformation of wing. The first one is to install aileron on the wing of the static test aircraft, and the wing is applied with matching load to carry out the test. This scheme directly simulates the real state, but it will affect the test cycle of static test aircraft. The second is to carry out aileron test on the bench, and design the support device with adjustable joint to simulate the wing deformation The scheme will not affect the whole test cycle, but it has higher requirements for the design of support device[7].

In this paper, the aileron supported on the wing was studied. In the test, if the static test aircraft is applied with matching load, the aileron assessment can be realized accurately, but there are too many loading points, resulting in a waste of resources. The aileron is installed near the wing tip on the outside of the wing. Only the deformation of the outer wing will affect the aileron strain distribution, while the deformation of the inner wing will only cause the aileron to produce rigid body displacement (Fig.4), and the influence on the aileron strain can be ignored. Therefore, in the test, it is not necessary to apply matching load on the whole aircraft, only the matching load on the outer wing and the trim load on the inner wing to prevent excessive wing deformation.

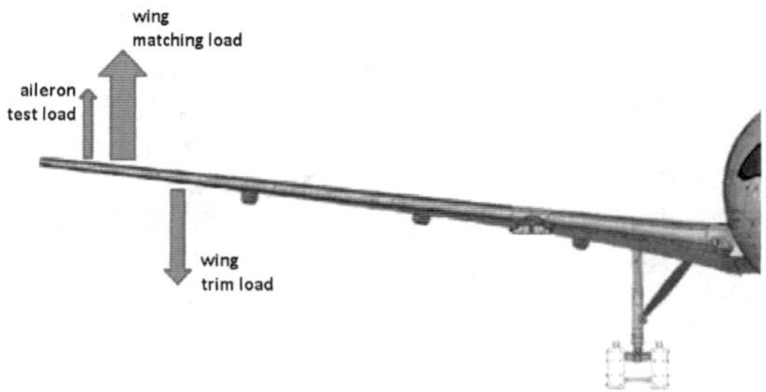

Figure 4. Test support scheme

4. Test Load

The determination of load is very important for the test whether it can meet the test requirements. In this section, the selection of aileron test load, load conversion and optimization were studied, the load screening principle and load processing steps were given, and the aileron test load design method was formed.

A. Load screening

In the flight load of large aircraft, there are dozens of serious load cases of aileron downward deflection. It is necessary to select representative dangerous case which can basically cover other cases. Because the strain distribution of aileron is affected by wing deformation, it is necessary to consider the serious situation after the superposition of aerodynamic load and coordinated deformation.

Aileron test load was selected according to the principle of "minimum M.S.". According to the critical conditions corresponding to the minimum safety margin of each main structure and main connection of aileron, the dangerous cases can be screened, and the cases can be combined according to the specific safety margin[8].

According to the internal force results of aileron finite element method and strength calculation method, the safety margin of aileron composite wing structure, joints and connections can be obtained. According to the safety margin, the preliminary screening of load dangerous cases was completed, and the strain distribution curve shows that the selected dangerous conditions can also achieve the assessment purpose in the secondary stress area of aileron structure. The dangerous load case of composite structure and metal joint may not be the same one, which can be evaluated separately or design a new test load according to the load cases.

B. Load processing

The original load is the aerodynamic distributed load, and the test load is the concentrated load at each loading point. After the dangerous load case was determined, the original load should be converted into the concentrated load at the loading point that can be used for test loading[9]. The aileron was divided into several stations according to the rib. When the rib spacing is too large, a virtual station can be added, as shown in Fig.5. Loading points were set at both ends of the rib station, the front loading point is the intersection point of rib and beam, and the rear loading point is the intersection point of rib and trailing edge.

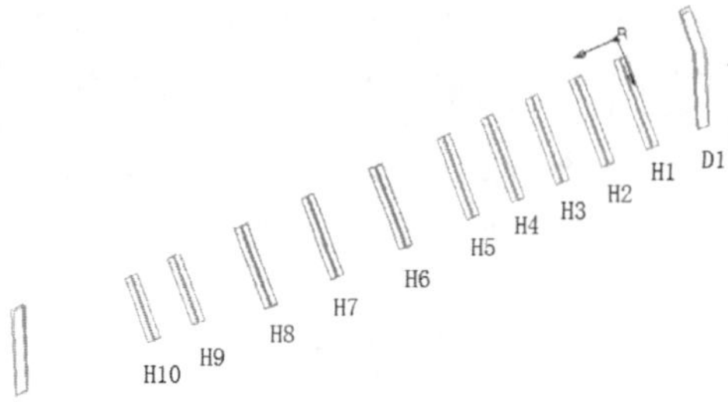

Figure 5. Rib stations

The load processing steps are as follows:

a) A local load coordinate system was established (Fig.6). The origin of the coordinate system is located in the center of the first hinge joint in the inner side.The X-axis direction is from the origin to the tail edge of the aileron,Y-axis is the direction of combined force of load under dangerous working conditions, the Z-axis is from the origin to the aileron spanwise outside.

b) The aerodynamic distributed load in the aircraft coordinate system was transformed into the local load coordinate system.

c) The transformed distributed load was processed by profile, and the distributed load of adjacent stations was combined into concentrated load of station profile.

d) The load of the rib station was evenly distributed to the loading point according to the moment balance.

Figure 6. Test load processing

5. Damage Introduction

In the process of manufacturing and using, the composite parts on aircraft will have defects or damages. According to the requirements of CCAR-25 and the recommendation of AC20-107B, manufacturing defects and damage in the process of use should be introduced in the full-scale verification test of composite structures, both of which are essential in the verification[10].

Before the test, it is necessary to complete the damage threat assessment of aileron structure, which includes manufacturing defect threat assessment and operational damage threat assessment[11]. The tasks to be carried out in the threat assessment of manufacturing defects include: to collect the information about the location and size of the defects of the test pieces produced by aileron suppliers in the early stage; to collect the inspection capability of the NDT system of each aileron supplier for various manufacturing defects; to collect the requirements for manufacturing defects in the manufacturing acceptance documents of aileron suppliers.

Operational damage threat assessment mainly includes the determination of damage source, damage location and damage size. The potential impact damage sources of aileron composite structure include: tool drop, hail, tire debris, runway gravel, lightning, rotor blasting debris, bird strike and other conventional impacts, as well as serious collision between service vehicles and aircraft, abnormal flight overload, abnormal hard landing and other unconventional impact. In the test, only conventional impact is considered[12].

The introduction principles of embedded defects in full-scale test pieces of aircraft composite materials are as follows:

e) The embedded defects should be arranged in the parts prone to manufacturing defects, such as the fillet area of the beam edge strip.

f) The embedded manufacturing defects should be placed in the dangerous details of the structure and have no coupling effect with other types of damage.

g) The embedded manufacturing defects should be located at the thickness of the structure with large interlayer stress.

h) Due to the need to monitor the status of defects during the test, the location of embedded defects should be easy to detect.

The principles of introducing damage to the full-scale test pieces of aircraft composite materials in the later stage are as follows:

a) The damage introduced should be arranged at the position easy to impact or scratch.

b) The damage introduced should be arranged at the dangerous details of the structure and not coupled with other types of damage.

According to the damage threat assessment results, combined with the introduction principle of embedded defects and impact damage, the type, location, size of aileron embedded defects and the size and location of impact damage were determined. The damage arrangement of aileron skin is shown in Fig.7.

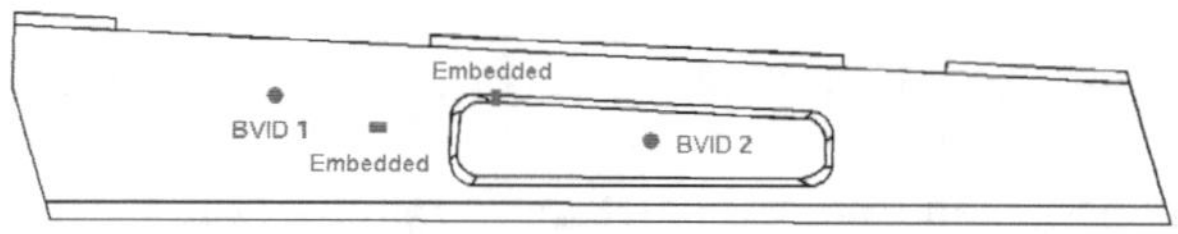

Figure 7. Aileron skin damage introduction

6. Aileron Test Measurement Scheme

Full scale test measurement includes strain measurement and displacement measurement. The strain measurement needs to reflect the overall strain distribution of the structure, and the displacement measurement results should be able to reflect the spanwise and tangential deformation of the structure[13].

The layout principles of aileron test strain gauge are as follows:

c) The arrangement of strain gauges should reflect the force transfer of the structure. Select some load transfer sections and arrange strain gauges to measure the strain distribution of the whole section and verify the finite element model. Strain gauges are arranged on the main force transmission path, especially on the single force transmission path structure. The load transferred on each force transmission path is calculated by strain measurement, and the finite element model is verified.

d) The overall strain value of aileron is small, and the verification strain gauges should be arranged at the position with larger strain value to facilitate data comparison.

e) The strain gauges used for verification should not be arranged in the parts with large stress gradient, and the parts with uniform cross-section should be selected for the arrangement of strain gauges.

f) Strain gauges should be arranged at key load transfer joints.

g) For the buckling skin under low load, the buckling condition was measured by pasting strain gauge.

The aileron displacement measuring points were arranged at the beam and rear edge of aileron, and 5 displacement measuring points were distributed on average. The specific number is adjusted according to the actual demand. Figure 8 shows the arrangement of aileron strain gauges and displacement measurement points.

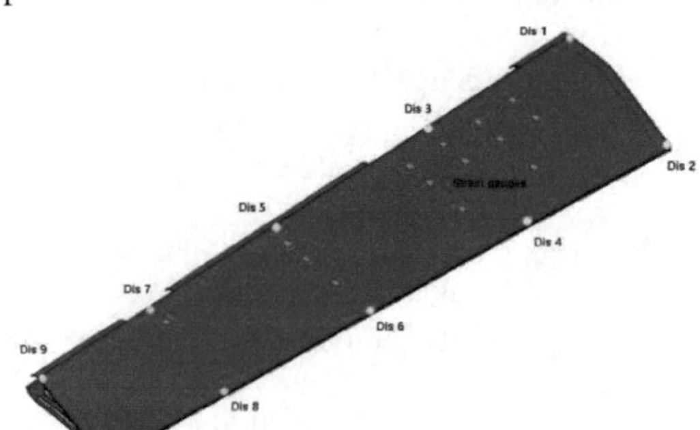

Figure 8. Test measurement scheme

7. Test Acceptable Criterion

The test acceptable criterion is determined according to the airworthiness clause, which is used to judge whether the test is acceptable or not. Before, during and after the test, corresponding inspection shall be conducted to ensure that the test meets the airworthiness requirements.

The test inspection contents are as follows:

h) Before the test, it is confirmed that there is no deformation of the assessment structure and the fasteners are in good condition.

i) During the test, the strain of the test structures and other bearing structures should be measured.

j) During the test, the load should be monitored in real time.

k) Monitoring the strain and displacement of key assessment parts.

l) Check whether the ultimate load holding time meets the requirement of 3 seconds.

m) After the test, check whether the structural parts are deformed and whether the fasteners are in good condition.

The acceptable criterion of aileron static test are as follows:

a) The ultimate load bearing time of the tested structure shall not be less than 3 seconds.

b) The tested metal structure has no crack or damage.

c) There is no harmful permanent deformation of the structure: a) there is no obvious permanent deformation of the structure through visual inspection; b) after the completion of each test, check that the aileron rotation is not stuck.

d) There is no fracture or damage of fasteners in the connection.

e) There is no harmful expansion of damage at the leading position of impact damage of composite structure.

f) There is no new damage in the composite structure.

8. Design Flow of Test Scheme

The test scheme of composite structure of large aircraft involves the research contents of test load, support mode, measurement method, damage introduction and other aspects, including various inputs such as airworthiness clauses, structural numerical model and analysis method[14-16]. In this paper, the whole process of test scheme design was studied, and the flow chart of test scheme design was formed, as shown in Figure 9.

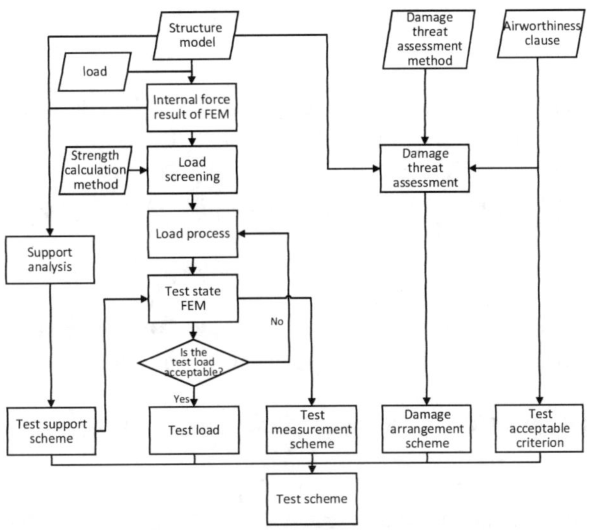

Figure 9. Test scheme design flow chart

9. Conclusion

In this paper, the test scheme design technology of composite aileron for large aircraft was studied, the main conclusions are as follows:

1. The influence of wing deformation on aileron strain should be considered in the test. In this paper, a test support scheme based on static test aircraft was proposed.

2. The method of load screening and load treatment for composite aileron was put forward, and the transformation from aerodynamic distributed load to test concentrated load was realized.
3. The damage sources of aileron structure were determined by using reasonable damage threat assessment method, and the introduction principles of embedded defects and impact damage were given.
4. The layout principle of aileron structure measurement scheme and test acceptable criterion were proposed to ensure the airworthiness compliance of the test.
5. A set of test scheme design process meeting the airworthiness requirements was formed, which can be used as reference for static test of composite structures.
6. According to the test scheme, the aileron ultimate load static test was completed, the test loading process was stable, the results meet the requirements of the test qualification criteria, and the aileron structure assessment was realized, which shows that the test scheme is effective and reasonable, and meets the requirements of airworthiness provisions.

References

[1] Chen S J. Composite technology and large aircraft[J]. Acta Aeronautica Et Astronautica Sinica.2008, 29(3),605-610. (in Chinese)
[2] Zheng Xiaoling. Research of airworthiness specifications for composite structure[J]. Civil Aircraft Design & Research, 2017(1): 1-7. (in Chinese)
[3] FAA. AC20-107B Composite Aircraft Structure[S]. Washington: FAA, 2010.
[4] Civil aviation administration of China, Chinese civil aviation regulations: Airworthiness standards transport category airplanes: CCAR-25-R4[S]. Beijing: Civil Aviation Administration of China, 2011. (in Chinese)
[5] Liu X K, Liu B, Zhang J F. Study on static test technology for high-aspect ratio wing of full scale aircraft[J]. Engineering & Test, 2014, 54(1): 17-20. (in Chinese)
[6] Xue C J,Tan W ,NIE H. Investigation on the Testing Technology of Pylon Static Test for Civil Aircraft Engine[J]. Journal of Experimental Mechanics, 2011, 26(6): 735-742.(in Chinese)
[7] Liu Q L,Yi W,Xia F. The Determination of Support Scheme for Aircraft Static Strength Verification Test[J]. Aeronautical Science & Technology, 2012:32-35.(in Chinese)
[8] Chen B, Lu H L. Statics strength test of the main landing gear attachment structure of a type of civil aircraft[J]. Civil Aircraft Design&Research, 2013(3): 21-23. (in Chinese)
[9] He Z Q, Liu Y, Li Z J. Load design for full scale static test of slat on large civil aircraft[J]. Acta Aeronautica et Astronautica Sinica. 2019, 40(2): 522197. (in Chinese)
[10] Si Erjian.Full-Scale Fatigue Test of Commercial Airplane Structure. Civil Aircraft Design & Research, 2012(1):47-52. (in Chinese)
[11] SANATEI M, ONIPEDE O. Damage assessment of structures using static test data[J]. AIAA Journal, 1991,29(7):1174-1179.
[12] Mao Y Q, Yang F, Wang L. Static strength Load and Experimental Study of Composite Wing of M2 Aircraft ,Advances in Aeronautical Science and Engineering , 2019, 10(5): 644-649. (in Chinese)
[13] Zheng J J, Tang J Y, Wang B W. Static test technology for C919 full-scale aircraft structure[J]. Acta Aeronautica et Astronautica Sinica, 2019, 40(1): 522364. (in Chinese)
[14] Sun X S, QI P Q. Design and verification guide of strength and stiffness in civil aircraft structure[M]. Beijing: Aviation Industry Press, 2012. (in Chinese)
[15] Liu J. Static test method for the after fuselage structure of civil aircraft[J]. Civil Aircraft Design &Research, 2013(2): 10-12. (in Chinese)
[16] Li W P, Tan W, Xue C J. Static test and computational analysis for pylon of airliner engine[J]. Journal of Nanjing University of Aeronautics & astronautics, 2011, 43(6): 732-737.(in Chinese)

Aerodynamic Design of Wingtip Devices Using Far-field Drag Decomposition

Xiaolong HE[a,*], Zheng CHEN[b] and Jie TAN[c]

China Academy of Launch Vehicle Technology, Beijing, China

[a] hexlong@outlook.com; [b] chenzju@live.cn; [c] tanjie1985@163.com

Abstract: Computational Fluid Dynamics (CFD) has become widely used for civil aircraft. However, the commonly used wall integration method for drag prediction based on Reynolds-averaged Navier-Stokes (RANS) equations still has some shortcomings. Far-field drag decomposition (FDD) method was developed as an alternative to improve the drag prediction accuracy and to provide a powerful analysis tool. FDD is capable of decomposing the drag into viscous, shock, induced, and spurious drag components, which provides invaluable insight into the relationship between drag and flow phenomena. Despite all the strengths, FDD is seldom used for aerodynamic shape design and optimization. In this paper, FDD was used to design wingtip devices for a business aircraft. Five spiroid and one blended winglet were compared with baseline wing and tip-extension configuration. Results indicate that all drag components were quantified successfully and the trade-off between induced drag and other drag components was better explained. The spiroid winglet with new foil gives the lowest total drag and wave drag.

Keywords: far-field drag decomposition, computational fluid dynamics, spiroid winglet,

1. Introduction

Nowadays civil transport aircraft need to address both economic and environmental concerns, which are directly related to aerodynamic drag. To reduce drag, wingtip devices have been widely investigated and various kinds of wingtip devices were proposed[1]. Spiroid winglet concept was proposed by Gratzer[2] and has been applied on a few airplanes[3-5]. Spiroid winglet is chosen as the research subject of this paper because of its advantages at reducing drag and accelerating vortex dissipation.

The accuracy of drag prediction is very important for winglet design because drag is used directly as the design merit. However, the wall integration method based on the solution of Reynolds-averaged Navier-Stokes (RANS) equations heavily relies on the mesh quality to achieve high accuracy. Insufficient grid quality will introduce numerical error. On the other hand, induced drag is hard to be quantified using near-field method. Far-field drag decomposition(FDD) method was proposed to provide a better choice. Wall integration method is also called near-field method, in contrast to the so-called far-field method. Paparone[6] proposed an FDD method based on Taylor's series expansion. He also introduced viscous and shock detector functions to locate the integration area. His study used Maskell wake survey method to calculate induced drag. Destarac[7] developed another FDD method that has been adopted by many scholars. His study defined every drag component by volume or surface integration. As for induced drag computation, some investigators proposed Q-criteria[8] or vorticity confinement method[9] for higher accuracy and robustness. Gariepy[10] proposed a new axial velocity defect(AVD) formulation to fix the discontinuity issue of Destarac formula when the total pressure is lower than the undisturbed static pressure. Lanzetta[11] compared a few viscous area detector functions. Marongiu[12] proposed an FDD method based on Lamb vector which gives a strict definition of induced drag but lacks the definitions of viscous and shock drag. FDD method has been applied in aerodynamic optimization[13], engine thrust and drag analysis[14]. However, FDD is not a mature technique yet and rarely used for practical engineering problems. The advantage of calculating induced drag will make winglet design problem a good demonstration of FDD. With our previously developed FDD tool[15], we could have more insight into the mechanism of winglets reducing drag. In this paper, we perform a quantitative analysis of a few wingtip devices of different kinds. Five spiroid winglets and one blended winglet were generated with the same span constraint and compared using FDD. The main contribution of this paper is that FDD helps reveal the variation of each drag component of winglets.

The following sections are structured as follows. Section 2 gives a brief introduction to the FDD method. Section 3 shows the application FDD method in winglet design and Section 4 draws some conclusions.

2. Theory of Far-field Drag Decomposition

In this section, we give a brief introduction to the FDD method. First, we discuss the basic theory and definition of each drag component. Then we explain the area selection strategy.

a) Definition of Various Drag Components

The theoretical basis of FDD is momentum theory. FDD method calculates far-field drag and it is equal to near-field drag. The drag balance is shown in (1).

$$\begin{aligned} D &= D_{\mathrm{nf}} = D_{\mathrm{ff}} \\ D_{\mathrm{nf}} &= D_p + D_f \\ D_{\mathrm{ff}} &= D_v + D_w + D_i + D_{sp} \end{aligned} \tag{1}$$

where D is the total drag. D_p and D_f are the pressure drag and friction drag from surface integration respectively. D_v is viscous drag caused by boundary layer and other viscous effects. D_w is wave drag caused by shock. D_i refers to induced drag. D_{sp} represents spurious drag which is caused by numerical errors and artificial dissipation.

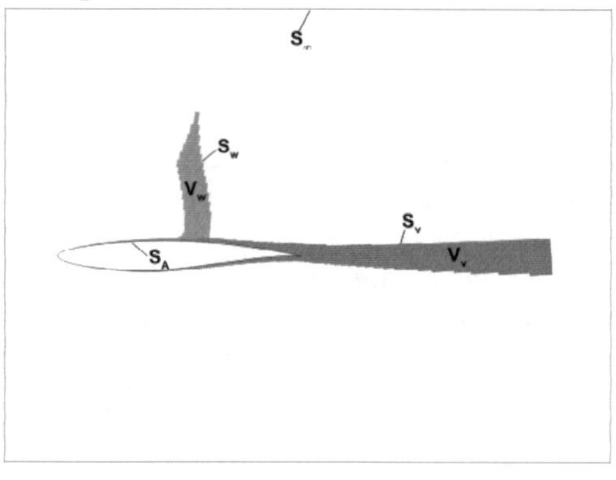

Figure 1. Example of flowfield and definition of viscous and shock area

Figure 1 shows that the flow field is decomposed into the viscous area, shock area, and other area. FDD method relates each drag component to a physical phenomenon which corresponds to a spatial area. For example, viscous drag is mostly generated in the boundary layer and viscous wake so it will be calculated by surface or volume integration in this area. S_A is the surface of aircraft and S_∞ is the far-field boundary. V_w and S_w represent shock control volume and surface. Similarly subscript v represents the viscous area. Likewise, we could derive the definition of all drag components, shown in (2).

$$
\begin{aligned}
D_v &= \oint\!\!\oint_{S_v} (\mathbf{f}_{vw}^* \cdot \mathbf{n})dS = -\oint\!\!\oint\!\!\oint_{V_v} \nabla \cdot \mathbf{f}_i^* dV + D_f = D_{vp} + D_f \\
D_w &= \oint\!\!\oint_{S_w} (\mathbf{f}_{vw}^* \cdot \mathbf{n})dS = -\oint\!\!\oint\!\!\oint_{V_w} \nabla \cdot \mathbf{f}_i^* dV \\
D_{sp} &= \oint\!\!\oint_{S_{sp}} (\mathbf{f}_{vw}^* \cdot \mathbf{n})dS = -\oint\!\!\oint\!\!\oint_{V_{sp}} \nabla \cdot \mathbf{f}_i^* dV \\
D_i &= \oint\!\!\oint_{S_v+S_w+S_{sp}} (\mathbf{f}_i^* \cdot \mathbf{n})dS = \oint\!\!\oint\!\!\oint_{V_v+V_w+V_{sp}} \nabla \cdot \mathbf{f}_i^* dV + D_p \\
\mathbf{f}_{vw}^* &= -\rho \Delta \bar{u} \mathbf{V} + \boldsymbol{\tau}_x \\
\mathbf{f}_i^* &= -\rho(u - u_\infty - \Delta \bar{u})\mathbf{V} - (p - p_\infty)\mathbf{i}
\end{aligned}
\tag{2}
$$

$\mathbf{f}$ is the integrator introduced by Destarac. $\mathbf{f}_{vw}^*$ is related to entropy caused drag and $\mathbf{f}_i^*$ is related to mechanical power loss. $\mathbf{i}$ is aligned with freestream direction. $\mathbf{n}$ is the normal vector pointing out cell face, S_v and S_w are the outer surface of the viscous area and shock area.. The definition of $\Delta\bar{u}$ is important for FDD method because it is used to define every drag component. Paparone, Destarac and Gariepy proposed three formulae for $\Delta\bar{u}$. This paper uses a hybrid method[15].

b) Area Selection Strategy

According to the theory of FDD, each drag component is related to a spatial area consisting of grid cells. So the detector function used to recognize these grid cells is critical for accurate drag prediction. Some scholars put effort into area detector functions to improve the robustness and accuracy of drag prediction results.

In this paper the viscous area function proposed by Paparone[6] is used.

$$F_{bl} = \frac{\mu_l + \mu_T}{\mu_l} \tag{3}$$

where μ_l and μ_T represent laminar viscosity and eddy viscosity. K_{bl} is a threshold value. When $F_{bl} > K_{bl} \cdot F_{bl\infty}$ it will be labeled as inside the viscous area. $F_{bl\infty}$ is freestream value.

Shock detector function is defined as:

$$F_{shock} = \frac{V \cdot \nabla p}{a\,|\nabla p|} \tag{4}$$

where a is the speed of sound, ∇p is the gradient of p. When $F_{shock} \geq 1.0$ it will be labeled as shock area.

Theoretically, the whole flow filed could be chosen as the integration area for induced drag. However, using an excessively large area will suffer from the transformation from induced drag to spurious drag. So a correction term is calculated by removing the correction integration area detected by the following function:

$$Q = \frac{1}{2}\left[|\Omega|^2 - |S|^2\right] > 0 \tag{5}$$

where Ω is vorticity tensor, S is strain tensor. $|\Omega|$ is Frobenius norm of Ω.

Using detector functions we can locate the shock wave but the drag result will not be accurate enough because of numerical oscillation. So an area expansion strategy is needed to get more reliable drag results[15].

3. Winglet Design and Analysis

In this section, FDD method is used in winglet design and analysis for a business airplane. Five spiroid winglets, one blended winglet, and one tip-extension configuration were generated and compared.

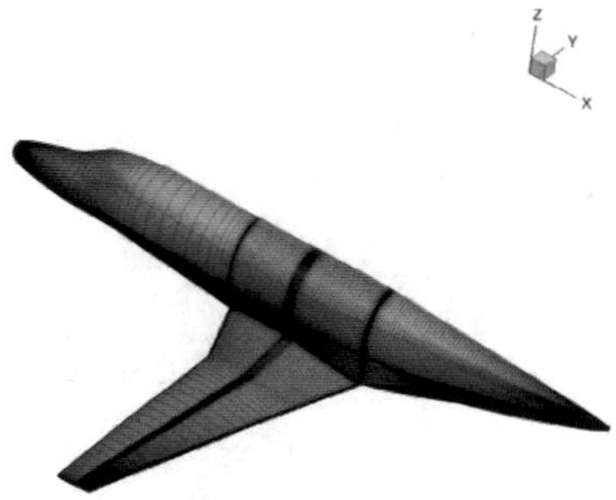

(a) Baseline shape

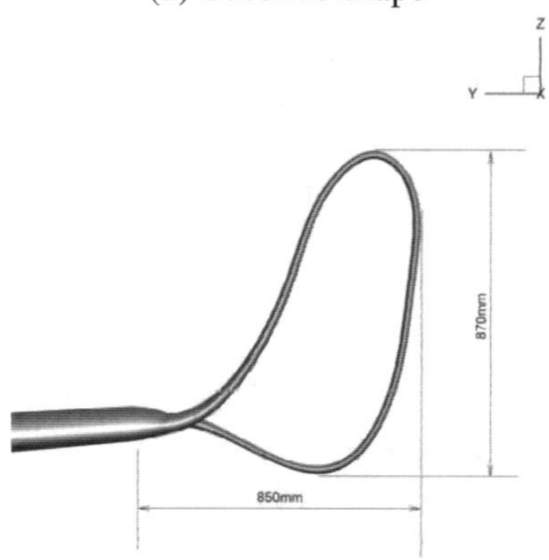

(b) winglet size

Figure 2. Baseline wing-body shape and winglet size

Figure 2(a) shows the baseline wing-body configuration and Figure 2(b) shows the baseline spiroid winglet. The reference area is $17m^2$ and Reynolds number based on the mean aerodynamic chord is 5×10^6. The cruise condition is at Ma=0.7 and C_L=0.45. A multiblock structured grid of 28 million cells is generated for the baseline winglet. The normal spacing of the first grid next to the wall is 1×10^{-6} and y+ value is lower than one. The same grid is used for baseline wing and blended winglet, which ensures the drag result is comparable. NACA0008 is chosen as the airfoil for winglet.

An in-house CFD code called TEAM is used in this paper. Jameson-Schmidt-Turkel spatial discretization scheme and Diagonally Dominant Alternating Direction Implicit (DDADI) time marching method were chosen. Multi-grid method, local time stepping, and implicit residual smoothing were also used to accelerate the convergence. Spalart-Allmaras(SA) turbulence model is also used for capturing turbulence effects.

Table 1 shows the near-field drag results of the baseline wing and all the winglets. The baseline shape is a business aircraft wing-body configuration. A tip-extension shape is generated by extending along leading edge and trailing edge with the tip airfoil. The blended winglet is also considered as a reference. The baseline spiroid winglet shape has the front winglet part at a higher position and the back part lower. All the other spiroid winglets were modified based on this baseline shape except the Spiroid_rev winglet. The Spiroid_rev winglet rotates in the opposite direction. The spiroid_foil shape was generated by replacing the winglet airfoil with NACA2408. The spiroid_height shape has enlarged winglet height of 990 millimeters. The spiroid_twist shape has 1.5 degrees larger twist angle for each control airfoil on the winglet.

In table 1 we can see that all the winglets show larger drag reduction than tipextension shape. Spiroid_foil and Spiroid_twist give lower drag than blended winglet. For these two winglets, most drag reduction is attributed to wing rather than winglet. Note that all the spiroid winglet carries larger lift and causes larger drag than blended winglet. The spiroid_foil shape has higher viscous drag than blended winglet but Spiroid_foil has lower induced drag. Summing all the drag components, Spiroid_foil is lower than blended winglet by 1.3 counts.

Table 2 compares the FDD results of all configurations. Adding winglets to the baseline wing, we can see that winglets lead to larger viscous drag. The Spiroid_rev gives largest viscous drag, which means it caused most viscous dissipation of flow energy along wingtip flow. Induced drag is also shown here, which indicates that main part of drag reduction is due to induced drag decrease. The best two winglets, Spiroid_foil and Spiroid_twist, both reduced induced drag most. Shockwave drag reduction also contributes to the drag reduction of winglet configuration. For example, Spiroid_foil shape gave 2.5 counts shock wave drag reduction relative to the baseline wing. Spiroid_foil shows the lowest drag 235.7 counts, which is 10.7 counts lower than baseline.

In figure 3 viscous area at wingtip is shown for each shape. In this figure, the blue surface is the winglet shape and the gray shaded area is the viscous area selected by viscous detector function. It could be seen that the tip-extension shape clearly reduces the size of the tip viscous area. All the winglets shape caused larger viscous area than baseline shape because of the enlarged wetted area. Blended winglet does not reduce the size of the viscous area compared with baseline. Spiroid_foil and Spiroid_twist gave larger viscous area than other spiroid winglets, which means faster vortex dissipation.

Table 1 Near-field drag results

Num	Name	C_D	wing				winglet			
			C_L	C_D	C_{Dp}	C_{Df}	C_L	C_D	C_{Dp}	C_{Df}
1	Baseline	0.024645	0.3890	0.01751	0.01277	0.00474	/	/	/	/
2	Tipextension	0.024115	0.3902	0.01714	0.01217	0.00497	/	/	/	/
3	Blended	0.023708	0.3834	0.01656	0.01182	0.00474	0.0077	0.00029	0.00001	0.00028
4	Spiroid_base	0.023831	0.3833	0.01648	0.01174	0.00474	0.0075	0.00047	0.00010	0.00037
5	Spiroid_rev	0.024017	0.3839	0.01670	0.01196	0.00474	0.0067	0.00044	0.00007	0.00037
6	Spiroid_foil	0.023573	0.3804	0.01613	0.01139	0.00474	0.0112	0.00065	0.00028	0.00037
7	Spiroid_height	0.023861	0.3834	0.01647	0.01173	0.00474	0.0075	0.00051	0.00011	0.00040
8	Spiroid_twist	0.023668	0.3814	0.01622	0.01148	0.00474	0.0100	0.00063	0.00026	0.00037

Table 2 Far-field drag results

Num	Name	C_L	D	D_v	D_w	D_i	D_{sp}
1	Baseline	0.45	0.024645	0.014617	0.000983	0.008980	0.000065
2	Tipextension	0.45	0.024115	0.014862	0.000856	0.008353	0.000044
3	Blended	0.45	0.023708	0.014963	0.000763	0.007961	0.000020
4	Spiroid_base	0.45	0.023831	0.015081	0.000788	0.007999	-0.000037
5	Spiroid_rev	0.45	0.024017	0.015224	0.000806	0.008082	-0.000096
6	Spiroid_foil	0.45	0.023573	0.015080	0.000738	0.007795	-0.000040
7	Spiroid_height	0.45	0.023861	0.015153	0.000789	0.007972	-0.000052
8	Spiroid_twist	0.45	0.023668	0.015080	0.000757	0.007853	-0.000021

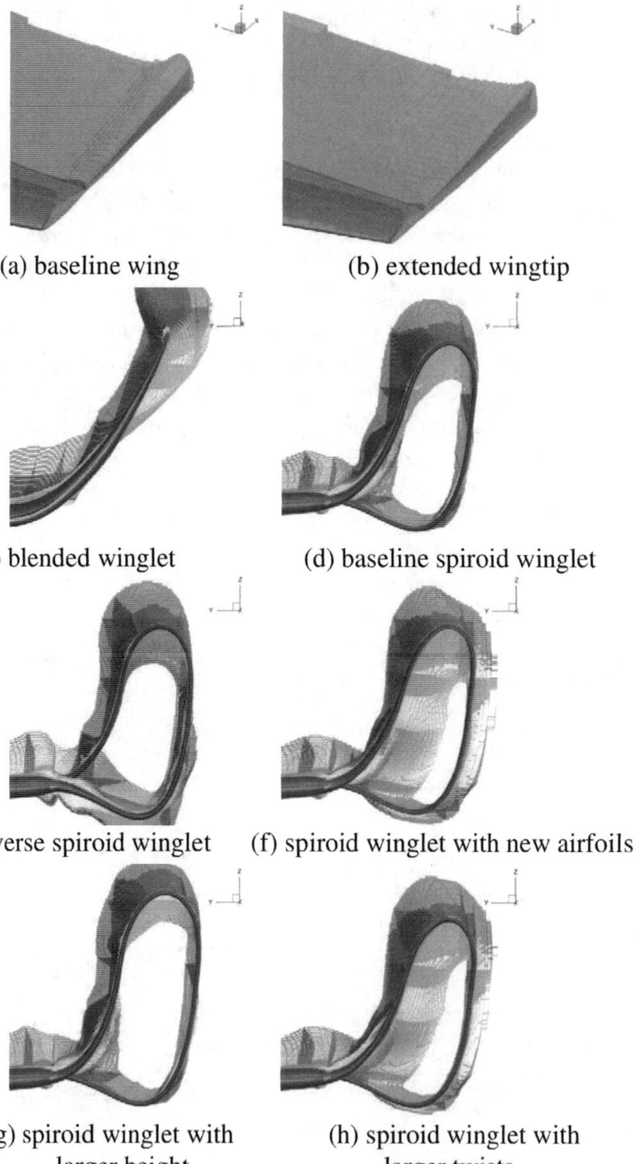

(a) baseline wing (b) extended wingtip

(c) blended winglet (d) baseline spiroid winglet

(e) reverse spiroid winglet (f) spiroid winglet with new airfoils

(g) spiroid winglet with larger height (h) spiroid winglet with larger twists

Figure 3. Viscous area visualization at the wingtip

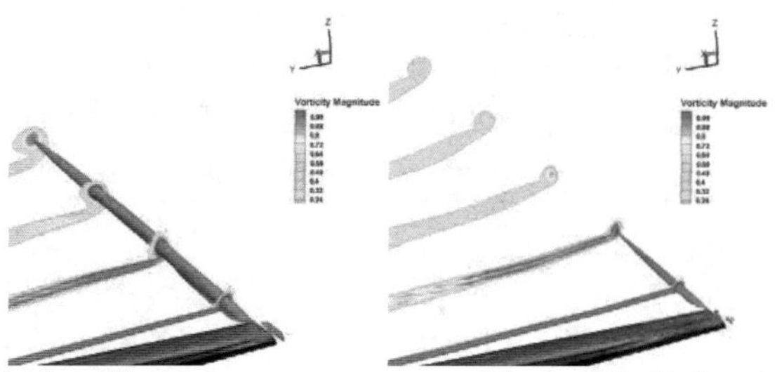

(a) baseline wing (b) extended wingtip

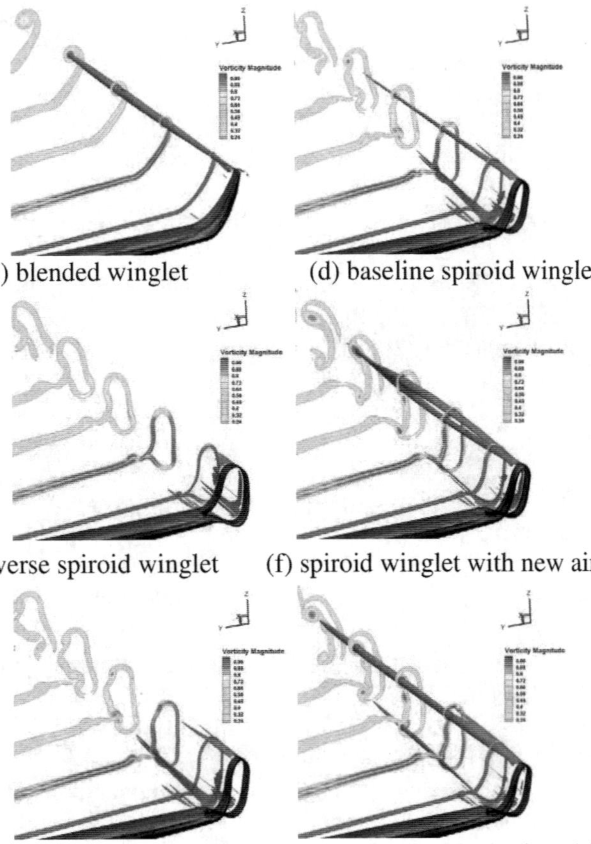

(c) blended winglet (d) baseline spiroid winglet

(e) reverse spiroid winglet (f) spiroid winglet with new airfoils

(g) spiroid winglet with larger height (h) spiroid winglet with larger twists

Figure 4 Viscous area visualization at the wingtip

Figure 4 compares the iso-surface of Q-criterion at the wingtip. It is shown that tip-extension gave smaller vortex size and blended winglet did not reduce the strength of the tip vortex. Baseline spiroid, reverse spiroid and higher spiroid winglet diffused the vortex after a few mean aerodynamic chords away from the wall. Spiroid_foil and Spiroid_twist still exhibited stronger vortex core strength than baseline spiroid winglet. These results show that the tip vortex strength is not directly related to induced drag. In other words, a lower strength of the tip vortex does not mean lower induced or total drag. The total drag is a balance between induced drag and other drag components.

4. Conclusion

In this paper, far-field drag decomposition method was used in the analysis of winglet and quantifying drag components. The results show that spiroid winglet could dissipate tip vortex faster than baseline shape and blended winglet. However, when reducing drag is the primary goal, the tip vortex strength may not decrease and even gets stronger. It could be seen that tip vortex strength does not relate to induced drag directly. Given the same lift, the circulation is fixed. So the only difference of different shape is how they distribute circulation. Even if the tip vortex is stronger, a configuration may still get lower drag. Using FDD, viscous drag, shock wave drag and induced drag could be quantified and more detailed analysis could be performed using these results. In our results, the shock wave drag reduction also contributes to the total drag reduction and the induced drag reduction is still the main contribution. The baseline spiroid winglet configuration shows better performance than reverse rotating spiroid winglet. Our approach heavily relies on the grid and solution quality, which could be improved in the future.

IOP Conf. Series: Materials Science and Engineering 1102 (2021) 012003 doi:10.1088/1757-899X/1102/1/012003

References

[1] Brocklehurst, A. and G. N. Barakos (2013). "A review of helicopter rotor blade tip shapes." Progress in Aerospace Sciences **56**: 35-74.

[2] Gratzer, L. B. (1992). Spiroid-tipped wing, US.

[3] Reneaux, J. (2004). "Overview on drag reduction technologies for civil transport aircraft." ONERA: Tire a Part(153): 1-18.

[4] Guerrero, J. E. and D. Maestro, et al. (2012). "Biomimetic spiroid winglets for lift and drag control." Comptes rendus - Mécanique 340 (1): 67-80.

[5] Guerrero J E, Maestro D, Bottaro A. Biomimetic spiroid winglets for lift and drag control[J]. Comptes rendus - Mécanique, 2012,340(1):67-80.

[6] Paparone, L. and R. Tognaccini (2003). "Computational fluid dynamics-based drag prediction and decomposition." AIAA journal 41 (9): 1647-1657.

[7] Destarac, D. and J. van der Vooren (2004). "Drag/thrust analysis of jet-propelled transonic transport aircraft; Definition of physical drag components." Aerospace Science and Technology 8 (6): 545-556.

[8] Bisson, F. and S. Nadarajah (2014). "Adjoint-based shape optimization for induced drag.".

[9] Snyder, T. and A. Povitsky (2014). "Far-field induced drag prediction using vorticity confinement technique." Journal of Aircraft 51 (6): 1953-1958.

[10] Gariepy, M. and J. Trepanier, et al. (2013). "Generalization of the far-field drag decomposition method to unsteady flows." AIAA Journal 51 (6): 1309-1319.

[11] Lanzetta, M. and B. Mele, et al. (2014). "Advances in aerodynamic drag extraction by far-field methods." Journal of Aircraft 52 (6): 1873-1886.

[12] Marongiu, C. and R. Tognaccini, et al. (2013). "Lift and lift-induced drag computation by lamb vector integration." AIAA journal 51 (6): 1420-1430.

[13] Yamazaki, W. and K. Matsushima, et al. (2008). "Aerodynamic design optimization using the drag-decomposition Method." AIAA Journal 46 (5): 1096-1106.

[14] Malouin, B. and J. Trépanier, et al. (2016). "Installation and interference drag decomposition via RANS far-field methods." Aerospace Science and Technology 54: 132-142.

[15] Lei Q., Xiaolong H., Yang S., Junqiang B., Li L. Far-field drag decomposition using hybrid formulas and vorticity based area sensors. Proceedings of the institution of mechanical engineers part G-journal of aerospace engineering. In press.

IOP Conf. Series: Materials Science and Engineering 1102 (2021) 012004 doi:10.1088/1757-899X/1102/1/012004

Influence of Basic Airfoil Layout on Aerodynamic Characteristics of Forward-swept Wing

Zhang Jun-yi[a], Su Xin-bing[b, *], Jiang Wen[c] and Zhao Xi-wei[d]

Air Force Engineering University, Xi'an, China

[a] akazikiitachi@163.com, [b]13572533404@163.com, [c] paimon1314@163.com, [d]18203618316@163.com

Abstract. When the swept angle of the wing varies, the parameters such as the aspect ratio of the forward-swept wing aircraft and the relative speed of the cross-section airfoil relative to the incoming flow will change, which will play a great influence on the aerodynamic characteristics of the wing. Based on this purpose, this paper explores the impact of the oblique angle of the basic airfoil on the aerodynamic performances of the forward-swept wing. Firstly, the wing models of the upright and oblique basic airfoil were established, meanwhile the one-equation Spalart-Allmaras turbulence model and CFD technique were used for numerical simulation calculation. The change law of the aerodynamic characteristics of the wing with the fundamental airfoil upright and tilted are compared and analyzed when the swept angle varies.

1. Introduction

The Morphing Aircraft can vary in standard layouts such as straight wing, forward-swept wing, and delta wing[1-2]. As the forward-swept angle varies, the wing aspect ratio and the relative thickness of the cross-section airfoil relative to the incoming flow will change accordingly, which plays a great influence on the aerodynamic performance of the forward-swept wing[3]. Therefore, in the process of morphing, it is of great importance to study the influence of fundamental airfoil oblique angle on the aerodynamic performances of the forward-swept wing[4-5].

This paper compares the aerodynamic performances of forward-swept wing with the swept angle at different oblique angles, and analyzes the flow mechanism by analyzing the flow field structure diagram. The intention is to find a reasonable oblique angle to arrange the ribs in order that the wing has good aerodynamic performances in both swept and straight wing conditions.

2. Forward-swept wing modeling

In this paper, NACA0006 airfoil with a thin symmetrical treatment is selected as the fundamental airfoil of the wing, and it is perpendicular to the leading edge of the wing when the airfoil is upright[6-7]. The wing configurations as the fundamental airfoil is upright and tilted are shown in figure 1.

Forward-swept wing models are established as the fundamental airfoil is upright and tilted, and they are imported into ICEM in order to use a structural grid to mesh them. The number of grids is controlled at about one million. The distances of the far-field boundary of the grid from up, down, front, left and right are about 10 times the average aerodynamic chord length, and the distance of the far-field boundary backward is about 15 times the average aerodynamic chord length, and there is no asymmetric surface. The distribution of the surface grid of the wing is shown in figure 2. The mesh is imported into FLUENT, and calculated under the condition of Mach numbers Ma = 0.2, as well as

Reynolds numbers Re = 1.1×10^6. A three-dimensional pressure-based implicit solver is used, and a transient algorithm is chosen, which can ensure the calculation accuracy when the angles of attack are high, and the energy equation is also concerned. The three-dimensional N-S equation and S-A turbulence model are used. The ideal gas is selected and the viscous flow model is solved by Sutherland's law[8]. The boundary condition is the pressure far-field boundary and the temperature is set to 288.15K.

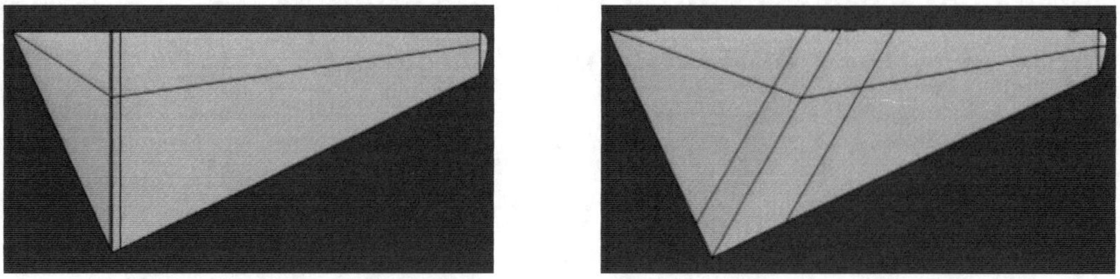

(a) Fundamental airfoil upright (b) Fundamental airfoil tilted

Figure 1. Airfoil configuration with basic airfoil upright and tilted

Figure 2. Grid distribution of wing surface

3. Aerodynamic performances analysis

3.1. Aerodynamic performances of the wing when the fundamental airfoil is upright

This article mainly analyzes the variety of the lift-drag performances and pitching-moment characteristics of the forward-swept wing with the forward-swept angle changing when the fundamental airfoil is upright. Figure 3 displays the curves of the drag, lift and pitching-moment coefficients of the forward-swept wing with the angle of attack changing when the forward-swept angles are 0 °, 20 °, 30 °, 40 °, and 50 °, respectively. In the graph of the pitching-moment coefficient curve, the gravity center of the wing is used as the reference point for the moment, but the position of the gravity center varies with the swept angle changing. Therefore, the moment reference points at different swept angles will vary accordingly.

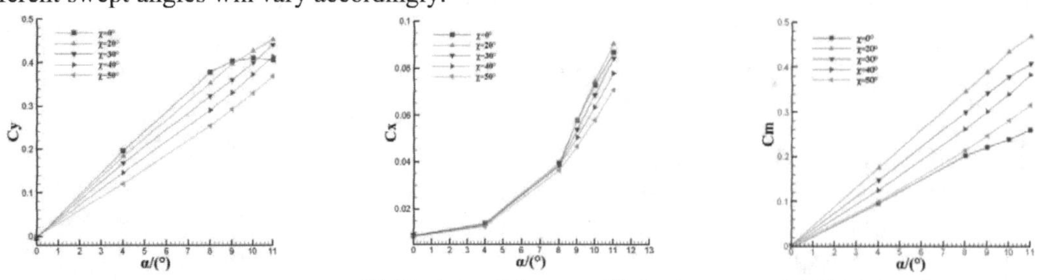

(a) Variety of lift coefficient with angle of attack (b) Variety of drag coefficient with angle of attack (c) Variety of moment coefficient with angle of attack

Figure 3. Impact of angle of attack on lift, drag, and moment coefficients at different swept angles

From the curve above, it is thus clear that for the lift performances, the coefficient gradually raises with the angle of attack raises, and at the condition of the medium and small angles of attack, the lift coefficient gradually falls down with the forward-swept angle raises,. When the forward-swept angle x = 0 °, the stall angle of attack is about 10 °, and the stall angle of attack of the wing in the forward-swept state is greater than α ⩽11 °. It can be seen that the raise of the swept angle can delay the wing stall.

With the raise of the angle of attack, the drag coefficient raises, and it raises more and more. When the swept angle is large, the drag coefficient curve is always lower than that as the swept angle is not large, which means that as the swept angle increases, the drag coefficient falls down. When the angles of attack are small and medium, the curves of different swept angles almost overlap, which indicates that the small swept angle has a small effect on the drag coefficient. As the angle of attack gradually raises, the curve spacing between different swept angles gradually increases, which indicates that the raise of the angle of attack makes the impact of the swept angle on the drag larger.

As the swept angle increases, the center of gravity and the aerodynamic center of the wing move forward. From the figure it can be concluded that as the angle of attack increases, the pitching-moment coefficient of the wing around the center of gravity gradually increases. When the angle of attack is α = 9° for straight wings, the raise in lift begins to slow down, so the raise in torque also gradually falls down, while when the angle of attack is constant, the torque coefficient firstly raises and finally falls down, which is caused by corresponding varieties in the gravity center and the aerodynamic center of the wing as the wing sweeps forward. As the swept angle raises, the distance between the center of gravity center and the aerodynamic center generally firstly raises and finally falls down, so that the maximum value is obtained when the swept angle is x= 20 °.

3.2. Aerodynamic performances of the wing when the basic airfoil is tilted

The aerodynamic performances of the wing is calculated when the fundamental airfoil is tilted at 30 ° in this paper. The relationship between lift, drag, moment coefficients and angle of attack at different forward-swept angles of the wing when the basic airfoil is tilted at 30 ° is shown in figure 4.

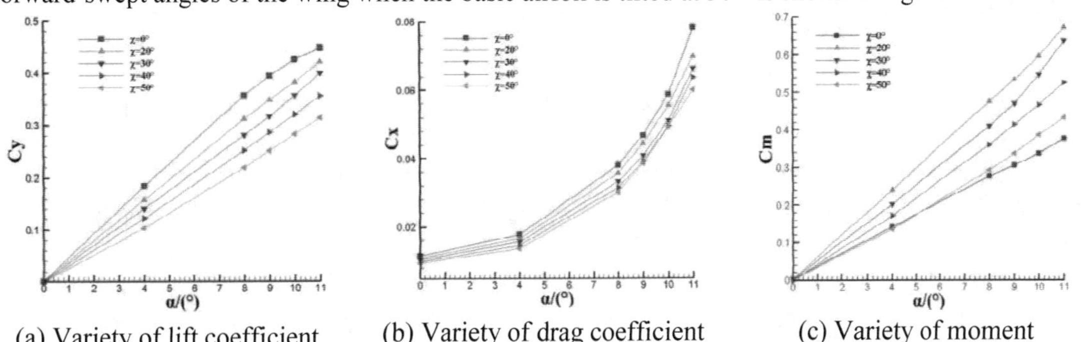

(a) Variety of lift coefficient with angle of attack (b) Variety of drag coefficient with angle of attack (c) Variety of moment coefficient with angle of attack

Figure 4. Impact of angle of attack on lift, drag, and moment coefficients at various swept angles

In general, figure 3 and figure 4 have the same general trend, and the varieties in lift, drag, and pitching-moment coefficients when the angle of attack are the same. When the angle of attack is the same, the parameters of lift, drag, and pitching-moment coefficients vary as the swept angle changing. In general, the lift coefficient of the wing with the fundamental airfoil tilted at various forward-swept angles is generally higher than that of the forward-swept wing with the fundamental airfoil upright. Comparatively speaking, the drag coefficient is significantly reduced, and the distance is enlarged, and the moment coefficient is also larger. This is because when the fundamental airfoil is tilted, the relative thickness and thickness distribution of the wing relative to the incoming flow will vary as the wing swept angle increases, which could also have varying degrees of impact on aerodynamic characteristics.

3.3. Aerodynamic performances of the wing when the fundamental airfoil is tilted

Figure 5 displays the pressure nephogram of the wing with the fundamental airfoil upright at various angles of attack under the condition of flat wing.

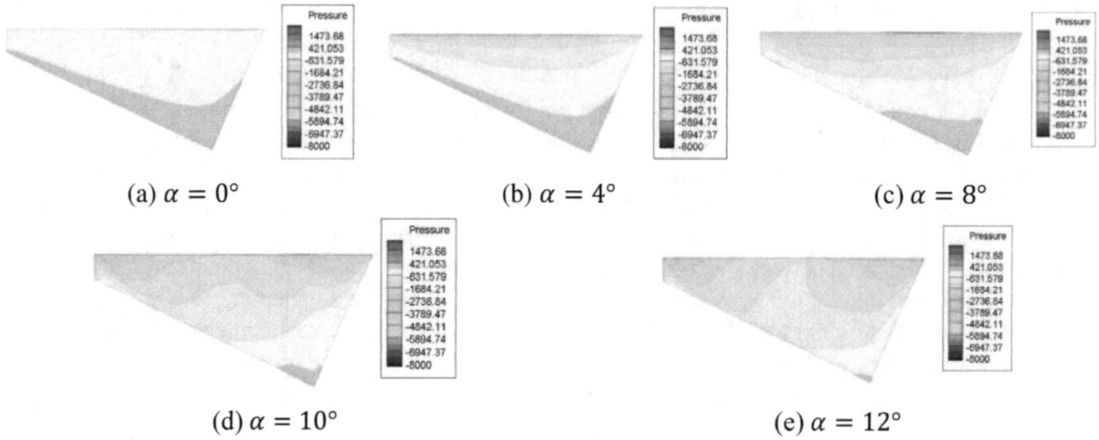

Figure 5. Pressure distribution contour of wing at various angles of attack when the basic airfoil is upright

From the contours above, it is thus clearly shown that when the angle of attack is not large, in the figure 5, when the $\alpha = 0°$, $\alpha = 4°$ and $\alpha = 8°$, the upper surface of the wing maintains the attached flow pattern, the boundary layer does not start to separate, the upper surface absolute pressure of the wing is small, and with the raise of angle of attack, the pressure has got a larger negative value. The negative pressure area of the leading edge gradually raises, and the suction impact raises. At this time, the lift coefficient raises due to the increase of angle of attack, and the growth relationship is similar to a linear curve. When the angle of attack raises, the negative pressure area of the leading edge of the wing expands further, the suction of the wing also raises, so the lift coefficient also raises. At this time, the upper surface of the wing begins to separate from the boundary layer, and the lift coefficient reaches the maximum value. When the angle of attack raises, the separation area of the boundary layer on the upper surface of the wing increases gradually, the vortex flow appears on the upper surface, the pressure on the upper surface appears uneven distribution, and the negative pressure area of the leading edge gradually disappears, the suction of the wing gradually falls down, the lift coefficient begins to drop sharply, and the wing appears obvious stall.

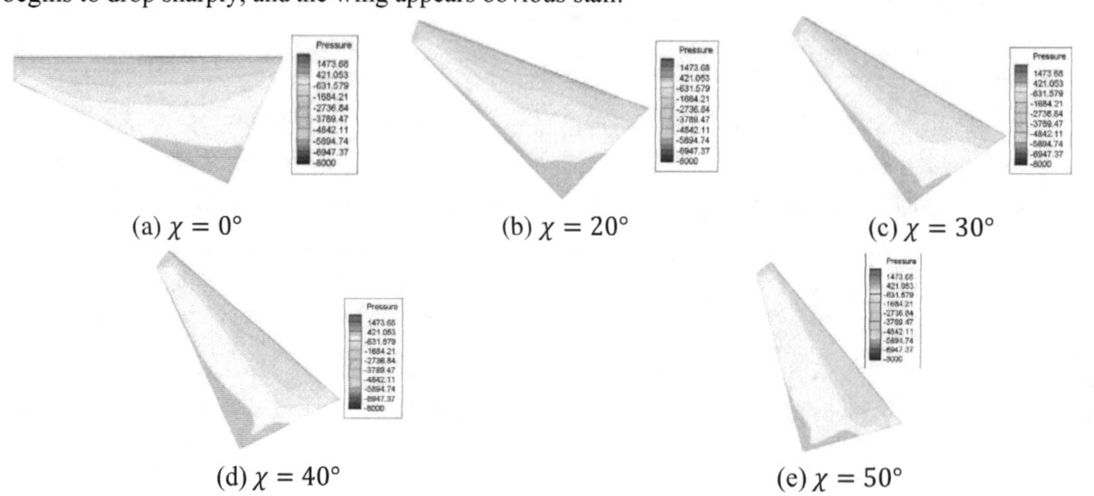

Figure 6. Pressure distribution nephogram of wing at various forward swept angles when the fundamental airfoil is upright and α=8°

Figure 6 displays the pressure distribution of the wing with the fundamental airfoil upright at different forward swept angles.

As figure 6 shows, when the sweep angle there is an obvious low pressure area on the upper surface of the leading edge of the wing, and the pressure distribution at this time is relatively uniform, then with the raise of the angle of attack, the area of the low pressure area of the leading edge of the wing gradually falls down, and the minimum pressure has a smaller negative value, the suction on the upper surface of the wing begins to fall down, and the lift coefficient gradually falls down; when the swept angle raises to 30° , the results show that the pressure distribution on the upper surface of the wing begins to be uneven, and then the lift coefficient falls down gradually.

4. Conclusion

Regarding the variety of the cross-section airfoil during the morphing process, the several following conclusions are obtained through numerical simulation calculation.

1) When the tilted angle of the fundamental airfoil is 0 °, the lift of the straight wing configuration is relatively high and the aerodynamic efficiency is good, but the wing stall is obvious. In the configuration of forward-swept wing, the stall angle of attack is large. The lift and drag are reduced after the forward-swept angle raises, and the torque has a larger negative absolute value.

2) When the tilted angle of the fundamental airfoil is 30 °, the impact of the forward-swept angle on the aerodynamic performances of the wing is the same as that of the forward-swept angle on the wing when the tilted angle is 0 °, while the difference is that a 30 ° inclined wing has higher lift and lower drag than a 0 ° tilted wing.

3) Regardless of whether the wing is in a straight or a forward-swept configuration, the tilted angle of the fundamental airfoil will significantly affect the aerodynamic performances of the wing, and the wing has good aerodynamic performances when the wing is tilted. Compared with the wing with the fundamental airfoil upright, when the fundamental wing is swept forward 30 °, the maximum lift coefficient is raised by 20% and the drag coefficient is reduced by 37%.

4) From the perspective of the aerodynamic performances of the straight wing and the forward-swept wing at various angles of attack, the wing has higher lift and lower drag when the fundamental airfoil is tilted, which means that it has better aerodynamic performances. Therefore, it is possible to use rib arrangement as the fundamental airfoil is tilted to design the wing as the wing of the variable forward-swept wing aircraft. In this case, the aerodynamic performances of the wing are all excellent during its morphing process.

References

[1] Neal D A, Good M G, Johnston C O, Robertshaw H H, Mason W H and Inman D J 2004 *45th AIAA/ASME/ASCE/AHS/ASC Structures, Structural Dynamics & Material Conf.* p 1727
[2] Secanell M, Suleman A. Gamboa P 2005 *46th AIAA/ASME/ASCE/AHS/ASC Structures, Structural Dynamics and Materials Conf.* p 1891
[3] Chen Q, Bai P, Ying WL, Leng JS, Liu ZQ 2010 *Acta Aerodynamica Sinica* vol 28 p 46
[4] Swaim RL, Newman BA 1985 *Journal of Guidance Control & Dynamics* vol 9 p 352
[5] Northrop Grumman Corporation 1999 *Aircraft with variable forward-swept wing* USA, US5984231
[6] Liu WF, Wang X 2009 *Acta Aeronautica ET Astronautica Sinica* vol 30 p 832
[7] Wang X, Su XB, Feng HY 2015 *Computer Integrated Manufacturing System* vol 32 p 54
[8] Chen TN 2000 *Applied fluid mechanics* (Beijing: Aviation Industry Press)

IOP Conf. Series: Materials Science and Engineering 1102 (2021) 012005 doi:10.1088/1757-899X/1102/1/012005

Operational Scenario Analysis Method for Civil Aircraft Based on Requirement Capture

He Yuwei[a,*], Sun Yuchen[b], Zhang Xinai[c], Dai Ximei[d] and Chen Dongsheng[e]

Requirements Engineering Department, Shanghai Aircraft Design and Research Institute, Shanghai, China

[a] heyuwei@comac.cc; [b]sunyuchen@comac.cc; [c]zhangxinai@comac.cc; [d]daiximei@comac.cc; [e]chendongsheng@comac.cc

Abstract: Scenario analysis is an accepted method in civil aircraft requirements capture and validation.This paper introduces the operation scenario analysis method for requirement capture, including stakeholder identification, interaction activity analysis, design requirement capture, etc.Based on SysML language, the modeling and analysis of operation scenarios were carried out.Taking the push/tow scenario of civil aircraft as an example, the structure diagram, use case diagram, activity diagram and requirement diagram are formed by modeling. The design requirements in this scenario are captured through the interaction analysis between the different stakeholders and the aircraft.
Keywords: Civil aircraft, System engineering, operational scenario, Requirement capture.

1. Introduction

As mentioned in SAE ARP 4754 and INCOSE[1],one of the difficulties in ensuring requirements integrity is that users do not always know what they do or do not want to do in the system, especially for new or novel features. There are many ways to get requirements from users. Early capture and prototyping of operational and maintenance scenarios is an example of how to capture requirements. It can be seen that scenario analysis is an accepted technical method in civil aircraft requirement acquisition and validation.

What kind of situation the aircraft will encounter in the whole operation process is the operational scenario of the aircraft, including the combination of various states and environments of the aircraft in the whole operation process. As the starting point and basic basis of aircraft design, the perfection of scenario definition is closely related to the correctness and completeness of top-level requirements, and also has a certain impact on safety analysis, test verification, airworthiness forensics and other work. If we can consider all the scenarios that the aircraft may encounter in the whole operation process at the beginning of aircraft design, it will save the development cost of the aircraft manufacturer and improve the efficiency of the aircraft after it is put into operation.

At the very beginning of aircraft development, the needs of stakeholders for aircraft products need to be determined. In addition, due to the long development cycle of aircraft products, the acquisition of requirements at the beginning of design also depends on the new operation and use requirements in the future. It is important to ensure that these requirements will apply in the future even if external conditions change, but uncertainty about the future makes it difficult to predict how future customer demand will develop[2].

One option to deal with future uncertainty is through scenario planning, which "represents the range of possible future developments" by creating multiple equally realistic descriptions of the future (scenarios). Scenarios capture future uncertainty in this way by considering a wide range of potential futures, and thus can form the basis for initiating robust design requirements. In this sense, scenarios, unlike prediction or trend extrapolation, do not represent natural states or predictions, but stipulate that systems that might operate in the future world must be reviewed. In fact, "scenario planning" has been used by aerospace companies for decades as a way to deal with future uncertainty.

The aircraft operational scenario is to put the expected aircraft in the entire air transport System at the very beginning of aircraft development and study the interaction between the aircraft and the System of System, such as the interaction between the aircraft and air traffic control, tower and other ground equipment. On the other hand, based on the concept of customer-oriented aircraft design, it is necessary to analyze the interaction between aircraft and external stakeholders and operating environment from the perspective of stakeholders using aircraft, which is operational scenario analysis.

2. Aircraft operational scenario

The original intention of Scenario, is derived from the drama, which refers to the gist of the plot. The general explanation can be understood as: a combination of events with certain relations. As a technology of requirements capture, scenario analysis refers to the method of obtaining requirements by placing the product to be developed in the scenario of its operation and analyzing its expected activity in the scenario. Scenario analysis is a creative way of thinking that reduces the likelihood that important requirements in the system will be ignored, while the activity of developing scenarios promotes communication within the organization[3].

Scenario is a concept originated from software engineering. Designers often use scenario to model the user of software system and the series of interactive activities between the system and the user. In activities such as early requirement extraction, it is easier for Stakeholders to use concrete examples than abstract intentions when communicating requirements. A scenario can act as a middle layer between an abstract conceptual layer and a concrete reality, and it can act as a means of putting requirements into context and articulating requirements clearly. With the help of scenarios, stakeholders can communicate their requirements through a series of specific, exemplary interactions.

In recent years, the application of scenario concept has drawn more and more attention from demand engineering, software engineering and other fields. People define different concepts of scenarios when they use them in different ways. A scenario is "an interactive activity that is likely to occur between multiple objects, and the activity has a goal". This interactive activity is a sequence of activities that takes place between objects. The sequence of activities describes a path where a goal is met or not met. The scenario contains the start and end states. A scenario can also refer to a specific example in a use case that represents a specific process in a use case in detail. Scenarios can also be combined with a variety of methods, such as viewpoints, cognitive science, etc.

In the design process of civil aircraft, operational scenario refers to a series of interactive activities and physical connections with temporal and logical relationships between aircraft and external environment (including relevant stakeholders, various external systems and external environment) for performing a certain task from the perspective of users using aircraft.

The operational scenario must be able to cover all stages of the flight, including takeoff, climb, cruise, descent, approach, and landing. The flight scenario must be able to cover significant flight environments, including flight planning, crew missions, Air traffic management (ATC), air traffic approval and Routing. The flight scenario must be able to cover typical flight weather, including visibility, turbulence, ice accumulation, thunderstorms, wind shear, crosswinds, etc. The flight scenario must be able to cover a variety of aircraft factors, including weight, balance, minimum equipment list, failures, abnormal events, etc. The flight scenario must be able to cover the necessary flight operations related to the crew mission. The flight scenario must reflect all fundamental functions and factors related to flight operations (CCAR 25.1523 Appendix D). The flight scenario must be able to reflect the aviation stakeholder factors that affect the flight safety to some extent.

Operational scenario analysis technology is used to describe the interaction between the aircraft and the operating environment, and can be used to identify aircraft functions and capture requirements. The developed aircraft can be placed in future operational scenarios, and the requirements can be captured by analyzing the expected activity of the aircraft in the scenario[4].

3. Operational scenario analysis method

a) Operational scenario analysis process

In the conceptual design phase of aircraft development, it is necessary to carry out the operational scenario identification and definition activities, and capture the complete design requirements through a comprehensive analysis of the scenario. For the identified operational scenario, the design requirements can be captured by analyzing the operational scenario.

Analysis methods for operational scenarios mainly include stakeholder identification, interaction analysis between aircraft products and stakeholders, requirement capture, etc., as follows:

- Stakeholder identification: in a certain scenario node, the personnel involved in aircraft operation include the stakeholders directly in contact with the aircraft, such as pilots, flight attendants, passengers, etc., and also include the stakeholders indirectly in contact with the aircraft, such as air traffic control personnel, dispatchers, etc.;
- Interaction activity analysis: In order to complete specific tasks, stakeholders need to carry out a series of interaction actions with the aircraft (including the transfer of resources, information, data, etc.), which are the operational behaviors of stakeholders. The description of the operation activity of the stakeholders should be in the language format of "verb/preposition + name", such as piloting the aircraft. In the process of describing the operation behavior, specify the external operation limitation/constraint in the activity description as required;
- Design requirements capture: Based on the above analysis, the requirements based on this scenario are captured. The captured requirements should be recombined or generalized according to category similarity.

b) Operational scenario modeling and analysis

Scenario concepts have been introduced into the field of engineering design to communicate design requirements. In a systematic design approach, the requirement list information can be refined and extended by creating scenarios during the task elaboration phase. A scenario describes the activities of a product (or system) and its participants in a narrative manner, including a series of interactions between users and the product (or system), such as which participants are likely to use the product, how to use it, when and where to use it, and the reaction of the product during use, etc. Essentially, a scenario usually tells of a series of operations imposed by stakeholders on a product (or system) and the events and actions (real or virtual) that the product (or system) responds to. In general, a scenario usually defines at least one set of stakeholder activities and at least one set of product (or system) actions. Much of this comes from stakeholder experience. Therefore, when using scenarios to express requirements, the description of specific instances makes it more convenient for stakeholders and designers to communicate requirements. In addition, scenarios can also be used by designers to model the interactions between products and their work environment.

A complete scenario should at least explicitly include the following characteristics: actors, time, place, actions/events, and scenario conditions (including preconditions and postconditions).

The "actor" is the person or other product or system that interacts with the product (or system) and is the subject of the action or event. It is important to note that scenarios describe not only the interactions between the product (or system) and the actors, but also the interactions between the actors or between the actors and the environment. In addition, actors often have different roles in a scenario, and roles indicate a specific type of actor. The terms "time" and "place" indicate when and where the scenario occurred, and are background elements that characterize the context of a general scenario. "Action/event" refers to an event in which an actor has an effect or occurs on a product (or system). Certain actions of one actor or product can cause changes in the state of others. Before and after the execution of the scenario, the state of each participant before and after constitutes the basic elements of the scenario conditions. Some conditions that must be met before the execution of the scenario (i.e., "pre-conditions") and those that should be met after the execution of a scenario (i.e., "post-conditions") are collectively referred to as "scenario conditions". It should be noted that the above several characteristic elements may not be included in every scenario. When one or more elements are omitted and the description and understanding of the scenario is not affected, it can be omitted.

Operational scenarios are widely used to ensure that the system (or set of systems) performs task requirements successfully. A operational scenario is a description of how the system should operate and interact with other systems. The scenario is described in a way that is easy to understand so that engineers can see how the various parts of the system work and interact, and verify that the system meets the needs and expectations of the user.

Describe a operational scenario that should target all operational patterns, task phases, and the sequence of critical activities for all users identified. Each scenario should appropriately include events, actions, information, and interactions to provide a full operational understanding of the system.

There are three main forms of scenario representation: natural language formal representation, formal language formal representation and semi-formal language formal representation. All three of the most simple and natural form is by people are known as natural language, in the form of a "story" analysis to observe people's task, the description method for all stakeholders are usually easy to understand, but natural language form is easy to produce ambiguity causes the description form in the real use fuzzy demand may be produced. Another way is to use formal languages such as CSS, CSP, Z to describe the scenario. This method requires relevant personnel to have high professional knowledge and technology, which is difficult for users without professional knowledge and technology to understand and master. Therefore, it will be very difficult to communicate with users to capture requirements. The third and semi-formal form of description is the graphical form, which is well observed, simple, and useful for users to understand, so this approach is better than the previous two for users and developers to understand.

In the field of aviation, scenarios can be divided into the following three expressions: diagram, table and model.

Modeling analysis of operational scenarios based on SysML[5] language. SysML(System Modeling Language) is a new Modeling Language proposed by INCOSE[1] (International Council on Systems Engineering) and OBJECT management organization OMG based on the reuse and extension of a subset of UML, and it is adopted as the standard Modeling Language for Systems Engineering.

SysML has a total of 9 graphics, including the following contents:

- Module definition: Said of elements such as modules and value type and the relationships between those elements, usually display system hierarchy tree as well as the classification tree;
- Internal module: used to specify the internal structure of a single module, display module, the relationship between the internal part and the interface connections between them;
- Use-case diagrams: Show the system of the interaction between the user and the system, the system is under the coordination of actors performing service by the black box view;
- Activity diagram: specify a behavior, mainly focus on control process, output and input through a series of moves into a process;
- Sequence diagram: How is part of the display module to interact through calls and asynchronous operation signal, state the action occurred over time and sequence of events;
- State machine diagram: Show a series of based on state of the system module, and respond to events, the possibility of conversion between state;
- Graph: Parameters said in one or more constraints (especially the equality and inequality) and how the property of the system bound, namely the constraint relation between modules;
- Package diagram: A model to show package contains the hierarchy of each organization;
- Demand figure: It means the relationship between the demand and the demand based on text, and meet, validation, and improve their other model elements.

- Combined with the above SysML model definition and characteristics, the operational scenario can be defined by use case diagram model and refined by establishing sequence diagram and activity diagram models.

During the modeling and analysis of operational scenarios, the structure diagram, use case diagram, activity diagram and requirement diagram of stakeholders should be established respectively.

- Structure diagram: Stakeholders in a operational scenario, which can be represented as SysML structure diagram. Should with the flow of information/data link (such as navigation, communication, etc.), resource flow/solid interface (fuel, gas interface, etc.), stakeholder-aircraft interaction (units and aircraft maintenance and aircraft, etc.), environment influence on aircraft such as the main line for reference, all to identify the stakeholders involved in the scenario, such as tower, air traffic control, satellite and radar, etc.;
- Use case diagram: Based on the above information identification system boundary (inside the boundary is the aircraft system, outside the boundary is the environment and stakeholders), use case diagram to simply describe the interface relationship between the aircraft product and external stakeholders; Stakeholder use case diagram model, as shown in Figure 1;
- Activity analysis : During the model expression of Activity analysis diagram, independent swimlanes should be set up for the aircraft and stakeholders (such as pilots, flight attendants, passengers, air traffic controllers, dispatchers, air traffic service system, navigation system, communication system, etc.), which should be expressed in the form of flow chart.Using the line to express the sequential workflow and interaction relationship; At the same time, the key indicators/parameter variables/decision points that affect the performance of the scenario are considered to set logic/trigger instructions, as well as operation restrictions/constraints.The model diagram of the activity diagram is shown in figure 2.
- Demand diagram: Design requirements can be captured and represented by demand diagram through stakeholder-aircraft interaction analysis under a certain operational scenario combined with model development experience, etc.

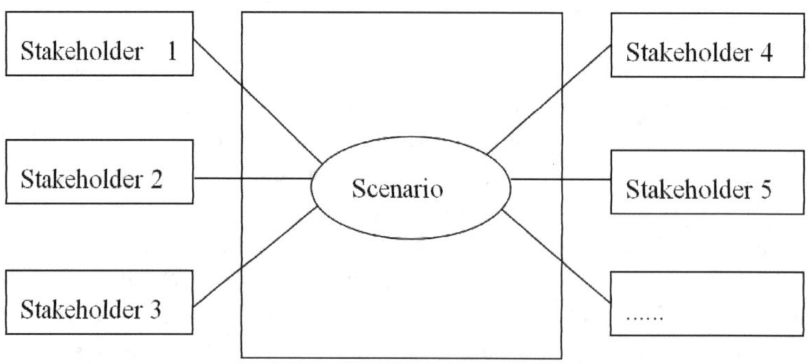

Figure 1. stakeholder use case diagram model diagram

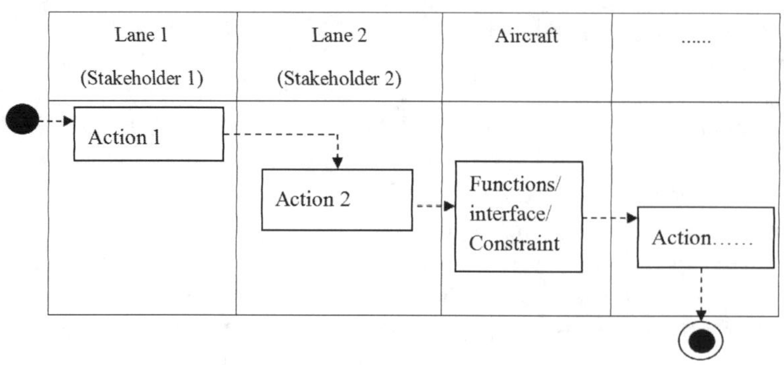

Figure 2. Activity analysis diagram model diagram

4. Operational scenario analysis example

This section describes an operational scenario analysis of a civilian aircraft for requirements capture, using aircraft push/tow as an example.

The aircraft push/tow scenario is a common operational scenario. This scenario mainly refers to the stage when the aircraft is in normal environment and normal operation in general operation, and when it is in normal environment and normal operation in general operation, the aircraft does not rely on its own power and USES ground equipment such as trailers to move in the airport activity area such as the corridor bridge, apron or parking space.

The stakeholders involved in this scenario include ground crew commander, tractor driver, service, maintenance, flight crew, air traffic control personnel (ATC), and flight attendants.

The main stakeholder-computer interaction activities include:

- a) Ground crew commander: Prepare for picking up the aircraft, direct the tractor to pre-position, direct the tractor to approach the aircraft at low speed, confirm that the tractor is connected to the aircraft, withdraw from the wheel gear, issue instructions for releasing the brakes, and direct the aircraft to push out/direct the tractor to pull out;
- b) Tractor driver: Pilot braking, confirm braking status, establish interphone, approach to the aircraft, put down and open the traction bracket, tow the aircraft/push the aircraft out;
- c) Service: Connect the tow bar to the aircraft, confirm that the brake is released, the wheel block is removed, and the front wheel turns and contacts the lock;
- d) Maintenance: Check the boarding bridge, evacuate the ground equipment, check the balance of fuel tank, check the compression amount of brake pressure damping pillar, and install the ground safety pin; Close the hatch cover, confirm the external lighting is normal, connect the internal communication;
- e) Flight crew: Apply for push/tow, set up external lighting, inform maintenance, prepare push/tow;
- f) ATC: Push/tow allowed;
- g) Steward: Direct passengers to board, close the gate.

The main stakeholder-aircraft interaction activities are shown in figure 5. According to the above stakeholder-aircraft interaction analysis, the aircraft should have the functions of providing external lighting, providing interphone, connecting with the tractor, wheel gear removal, brake release, front wheel turn contact locking, being pulled/pushed to the specified position and so on.

a) Modeling analysis

Based on SysML language, we use Magic Draw software to conduct the push/tow scenario analysis. Structure diagram is shown in figure 3, use case diagram is shown in figure 4 and scenario activity diagram is shown in figure 5.

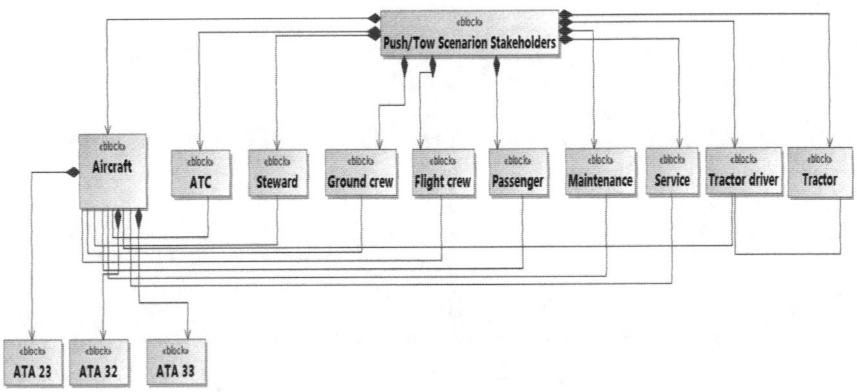

Figure 3. Stakeholder structure diagram

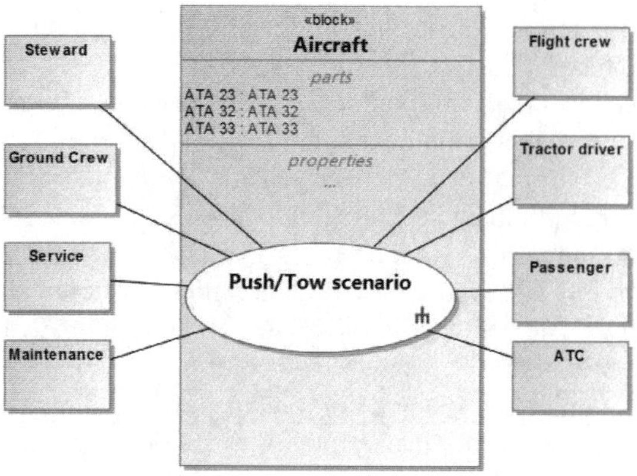

Figure 4. Use case diagram

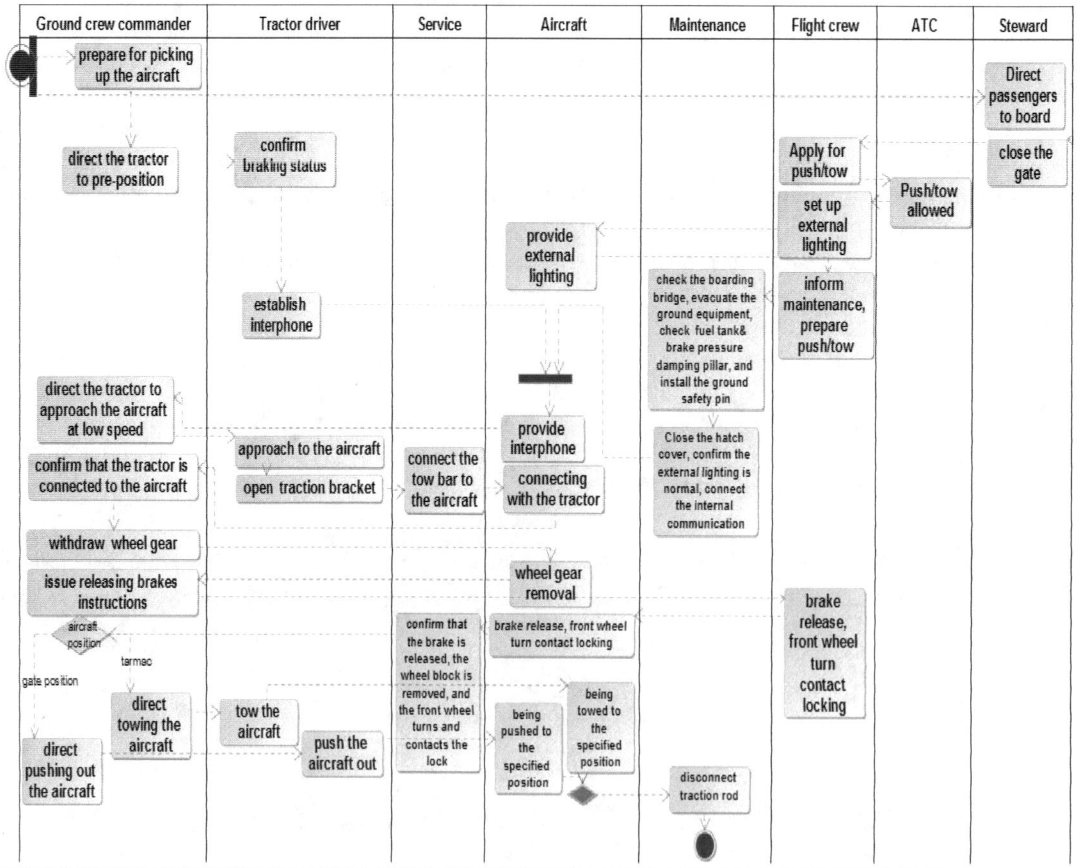

Figure 5. Activity diagram

b) Push/tow scenario design requirements capture

The main stakeholder-aircraft interaction activity in push/tow scenario is shown in figure 5.Based on the stakeholder-aircraft interaction analysis of the above aircraft push/tow scenario, the following design requirements are preliminarily captured:

- Aircraft shall provide sufficient means of internal and external communication;

- The aircraft lighting system shall provide external lighting to ensure the safety of the aircraft during takeoff, landing and navigation.
- The aircraft should have an interphone system to support real-time communication between the cockpit and ground service personnel;
- During aircraft installation and maintenance, the nose landing gear and cabin door shall be locked with safety pins;
- The cockpit of a civil aircraft shall provide sufficient field of view for the crew to observe conditions on the ground outside the aircraft;
- Under the maximum allowable towing weight of the aircraft, the civil aircraft shall be towed or propelled by a towed connected to the nose landing gear strut, and the nose landing gear shall have a joint connected to the towing accessories through which the aircraft shall be towed;
- The aircraft shall have braking function.

Acknowledgment

In the early stage of civil aircraft development, it is necessary to carry out aircraft operational scenario research. Through comprehensive scenario analysis, complete design requirements are captured, which is of great significance to improve the development level of new aircraft in China, improve aircraft design efficiency, and improve aircraft operation efficiency. In this paper, the operational scenario analysis method for civil aircraft based on requirements acquisition is proposed. Through the operational scenario analysis, the design requirements under this scenario are captured, which provides a feasible method for design requirements acquisition.

References

[1] International Association of Systems Engineering, INCOSE. Systems Engineering Manual [M]. Zhang Xinguo, trans. Beijing: Machinery Industry Press, 2014.
[2] A Guidelines for Development of Civil Aircraft and System [S]. The Engineering Society for Advancing Mobility Land Sea Air and Space, 2010. 5.
[3] Xie Ling, Fang Junwei, Xuzhou, et al. Research on aircraft requirements acquisition and confirmation Based on functional scenario analysis [J]. 2015,13 (18): 83-84;
[4] Deng xingmin, zhang huiyuan, li jianren, Yang chao. Application of requirement capture method based on operational scenario in turret [J]. Journal of northwest university of technology, 2017, 35(S1):88-92.
[5] Ni Zhongjian, Zhang Yan, et al. Application research of SysML system Design method [J]. Avionics technology, 2011,42 (1):18-23.

IOP Conf. Series: Materials Science and Engineering 1102 (2021) 012006 doi:10.1088/1757-899X/1102/1/012006

A Review of Advanced High-Speed Rotorcraft

Xufei Yan[1,a], Bin Lou[1,b], Anhuan Xie[1,c], Lingkai Chen[1,d] and Dan Zhang[1,e] *

[1] Zhejiang Lab, Hangzhou, Zhejiang, 310000, China

[a]yanxufei@zhejianglab.com; [b]loubin@zhejianglab.com; [c]xieanhuan@zhejianglab.com; [d]clk@zhejianglab.com; [e]dzhang@zhejianglab.com

Abstract. Urban air transportation (private aircraft, air taxi, etc.) and all kinds of special flight missions (aerial sightseeing, logistics transportation, disaster relief, anti-terrorism and anti-riot etc.) put forward high requirements for the flight performance of future urban aircraft. It not only demands high-efficiency hovering performance, but also demands the capability of high-speed cruising as well as high safety, clean energy and other characteristics. The maximum flight speed of conventional rotorcraft (such as helicopter, multi-rotor UAV, etc.) is limited by the fluid compressibility of advancing rotor blade and the airflow separation of retreating rotor blade, which makes further improvement of the flight speed more difficult. In order to overcome the fluid compressibility of rotor advancing side and stall of rotor retreating side encountered under high-speed, researchers have proposed many new concepts of advanced rotorcraft. This article sorts and analyses the current advanced VTOL (Vertical Take-Off and Landing) rotorcrafts with new configurations, and attempts to point out the feasible research direction in the future.

1. Introduction

Conventional rotorcraft (such as helicopter, multi rotor UAV, etc.) has efficient performance in vertical take-off and landing, hover, low-altitude and low-speed flight, as well as the unique capabilities of back and side flight. However, due to the asymmetry of the flow between the left and the right sides of rotor in forward flight, the maximum flight speed is limited by the fluid compressibility of the advancing rotor blade and the flow separation characteristics of the retreating rotor blade [1]. According to the research and analysis of rotorcraft aeromechanics, it is very challenging to design a new type of rotorcraft, not only to meet the high-efficiency hovering performance like helicopter, but also to meet the high-speed cruise capability like fixed wing aircraft, and under reasonable cost. The key factor in the design of such aircraft is rotor disc load. Figure 1 shows the relationship between the hovering efficiency and disc load of different types of VTOL vehicles [2]. It can be seen that the aircraft with lower disk load has better hovering efficiency (such as conventional helicopter, multi rotor UAV, etc.), with the increase of disk load, the hovering efficiency gradually decreases, but the maximum flight speed increases (such as F35B Fighter, Harrier Fighter, etc.). It can be seen that the VTOL performance and high-speed flight capability are contradictory. The design of advanced rotorcraft should consider the trade-off between flight speed, hovering efficiency, manufacturing complexity and cost.

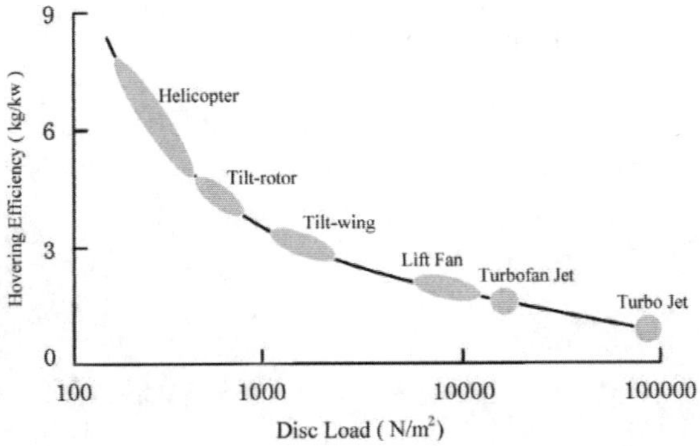

Figure 1. Relationship between hovering efficiency and disc load

For this reason, researchers have made a lot of explorations and attempts, and proposed various new configurations. After decades of development, demonstration, and experiments, the current high-speed rotorcraft mainly adopts three technical solutions: compound high-speed rotorcraft [3~6], rotor/wing conversion high-speed aircraft [7~9] and tilt-rotor (wing) aircraft [10~15]. This article mainly introduces and analyzes the related research of these three technical solutions, and attempts to point out development direction in the further.

2. Review of related work

2.1. Compound high-speed rotorcraft

2.1.1. Winged compound rotorcraft. In recent years, the winged compound rotorcraft that features the fusion of wing and multi-rotor power system has developed rapidly [16]. This kind of rotorcraft is usually equipped with a wing to provide lift together with the rotor at high speed, and the tail rotor of conventional helicopter are replaced by the propeller system with adjustable thrust, as shown in Figure 2.

The core idea of the technical solution of winged compound rotorcraft is: to unload the main rotor through the combination of wing and multi-rotor power system. In hover and low-speed flight, the aircraft operates in helicopter mode. When the flight speed increases to a certain level, the rotor speed will reduce with the rotor unloaded and the wing gradually loaded. At this moment the forward thrust is borne by the propeller system [17]. When the rotor speed is reduced to half, the main rotor generally needs to bear merely 20% of the flight weight, this operation mode is mainly used for high-speed flight [18]. The decrease of the rotor speed and the angle of attack of the blade profile delay the shock stall of the advancing rotor blade and the airflow separation of the retreating rotor blade respectively. The winged compound helicopter not only increases the flight speed and expands the flight envelope, but also greatly reduces the vibration level and improves the manoeuvrability. Figure 2a shows the concept aircraft eCRM-002 proposed by Uber [5]. During vertical take-off and landing, the aircraft provides lift through the coaxial co-rotating rotors of the distributed electric propulsion (DEP) system (the coaxial co-rotating rotors can reduce noise and improve aerodynamic efficiency); when flying at high speed, the rotors of the DEP system are unloaded, the wings are gradually loaded, and forward flight thrust is provided by two propellers located at both tips of the wing. Figure 2b shows the X-49A [19] "Speed Eagle" compound high-speed helicopter, which is modified from Sikorsky's YSH-60F helicopter. It is equipped with a wing to unload the main rotor, and a ducted fan vector propulsion system to balance the negative torque of the rotor and conduct heading control.

More representative models include Eurocopter's X3 high-speed compound helicopter (Figure 2c, in the certification stage [3]), Boeing's PAV (Figure 2d, flight test of scaled prototype [20]), and AutoFlightX's V600 (Figure 2e, prototype display [21]), Geely Automobile's TF-2A (Figure 2f, flight test of scaled prototype [22]), Kitty Hawk Corp.'s Cora (Figure 2g, prototype flight test [23]), and Beta Technologies' Alia eVTOL (picture 2.1h, prototype flight test [24]).

(a) Uber eCRM-002 (b) X-49A
(c) X3 (d) Boeing PAV
(e) AutoFlightX V600 (f) Geely TF-2A
(g) Cora (h) Alia eVTOL

Figure 2. Hybrid compound helicopter

2.1.2. ABC rotorcraft. As early as the mid-1960s, Sikorsky Aircraft Company took the lead in researching an "Advancing Blade Concept" (ABC) rotor [25, 26], aiming to break through the limitations of the working mechanism of traditional helicopter rotor. The design feature of the ABC rotor system is the application of two coaxial rigid rotors rotating in opposite direction, which allow the lift to be moved to the advancing side of the disc, thus the retreating side of the disc can be offloaded at high speeds and therefore avoid blade stalling. A technology demonstrator ABC was tested in the 1970's (XH-59A, Figure 3a), it is able to achieved speed of 240 knots, but with high vibration and hub drag [26]. Recently, the ABC rotor system has been revisited and the design has been improved through the use of advanced aerofoil sections and active vibration control. Due to these improvements as well as the pusher propeller providing an extra component of horizontal thrust, the Sikorsky X2 (Figure 3b) is able to reach speeds of 250 kt, which is 60% higher than the conventional helicopter [4].

During cruising of Sikorsky X2, the rotor speed reduces to delay the compressibility of airflow of the advancing rotor blade, and most of the engine power is transmitted to the propeller, which is the key to high-speed flight. In order to avoid interference with the operation of the coaxial rigid rotors, the vertical tail is inverted and the rudder is designed to improve the stability and maneuverability of the helicopter. The long horizontal tail wing bends the wing tip down to form two auxiliary vertical tails.

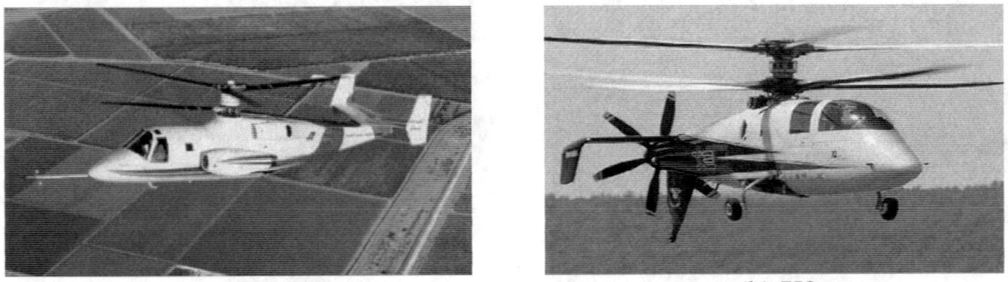

(a) XH-59A (b) X2

Figure 3. Coaxial compound helicopter

The compound solution of coaxial rigid rotor and propulsion system presents new challenges to rotor structure, material and assembly technology. With the increase of rotor size, the stiffness of rotor blade will decrease. Therefore, the ABC rotor system is not yet applicable for large rotorcraft.

2.1.3. Development direction of compound solution. The compound high-speed rotorcraft is as efficient as that of a conventional helicopter in hover. However, when flying at high speeds, the rotor speed is reduced to delay the compressibility of airflow of the advancing rotor blade, thus the reverse flow region is very large. When the advance ratio is greater than 1, the retreating blades will be mostly in the reverse flow region, which makes the aerodynamic environment very harsh at this time [27]. In addition, due to the configuration of wing and propulsion system, the structural weight increases, and the aerodynamic interference between the wing and the rotor is worse. Therefore, the technology of the compound high-speed rotorcraft needs further research and development, especially in solving the problems of aerodynamic interference and extra structural weight.

2.2. Rotor/wing conversion high-speed aircraft

The development of compound high-speed rotorcraft cannot avoid the asymmetric characteristics of the airflow generated by the rotor. The only way to eliminate asymmetric airflow in forward flight is to stop the rotor rotation, which is the core concept of the rotor/wing conversion high-speed aircraft [7]. This kind of aircraft can completely convert from helicopter mode into fixed-wing aircraft mode, which solves the problem of conventional rotorcraft flight speed limitation.

The stopped rotor / X-Wing rotorcraft proposed by NASA Lewis Research Centre can realize vertical / short takeoff and landing (VSTOL) and high-speed flight at Mach 0.85 [8]. The X-Wing rotorcraft has four rigid blades, which work like traditional rotors when taking off, landing and hovering in helicopter mode. In cruise flight, the clutch separates the engine shaft from the rotor and the rotor is fixed in an "X" shape. The rotor turns into a forward swept and a backward swept fixed wing with an angle of 45 degrees from the fuselage. The engine propels the aircraft by jetting backwards, as shown in Figure 4a. Boeing has developed the X50A "Canard Rotor/Wing" aircraft CR/W [9] (Canard Rotor/Wing), which applies tip jets to drive the blades. After reaching a certain speed level, the rotor stops and locks as a wing shape. During the transition process, the canard and horizontal tail bear the lift, and then the CR/W will fly forward at high speed in the fixed-wing aircraft mode, as shown in Figure 4b.

Compared with the X-Wing, the transition of CR/W flight is very convenient. The horizontal thrust is provided by conventional turbofan engines, which is a perfect combination of half helicopter and half aircraft.

The breakthrough and solution of key technologies such as the airfoil aerodynamic performance, blade control, rotor drive system and rotor/wing conversion technology will bring the development of the rotor/wing conversion aircraft to a higher stage. At present, the above technology is yet difficult and immature.

(a) X-Wing rotorcraft

(b) X50A CR/W

Figure 4. Rotor/wing conversion aircraft

2.3. Tilt-rotor/wing aircraft

The tilt-rotor/wing aircraft is equipped with one or more pairs of tiltable propeller/rotor assemblies [10~15]. This type of aircraft performs vertical take-off and landing, hovering and low-speed flight in helicopter mode. When the flight speed reaches a certain level, the rotors tilt to horizontal position along with the engine nacelles (or wing). At this time, the wing is gradually loaded, with the rotor unloaded. When the engine nacelles are tilted to horizontal position, the rotors become propulsion propellers, and the aircraft enters high-speed flight in a fixed-wing aircraft mode. In order to balance the hovering performance and high-speed flight performance, the negative twist angle of rotor blade is larger (between -10° of conventional rotor blade and -60° of propeller blade), thus its hovering efficiency is lower than that of conventional helicopter. Tilt-rotor / wing aircraft can be classified into three typical types: tilt-rotor, tilt-ducted fan and tilt-wing.

2.3.1. Tilt-rotor aircraft. Tilt-rotor aircraft has good vertical flight and high speed cruise performance. The propeller / rotor blade of tilt-rotor aircraft has a negative twist angle of -40°~ -50°, which can work in a large axial inflow state. In high flight speed, the wing provides lift for the aircraft, and the propeller / rotor is mainly used to overcome the profile drag, as shown in Fig. 5. Research has shown that a tilt-rotor aircraft with advanced technology of rotor airfoil, hub configuration, wing airfoil and fuselage structure weight can reach a maximum flight speed of 740km/h (400kts), twice that of conventional helicopters. The range of conventional helicopter rarely exceeds 1000 km, while the flight envelope of tilt-rotor aircraft exceeds that of helicopter and turboprop aircraft respectively. The operational range of a typical tilt-rotor aircraft V-22 is more than 1850 km. If two more transfer tanks are filled, the range can reach 3890 km.

In the 1980s, the verification prototype XV-15 tilt-rotor aircraft of Bell Helicopter [28] (Fig 5a) successfully verified the feasibility of the tilt-rotor program. In recent years, the V-22 Osprey tilt-rotor aircraft (Fig. 5b) which has been in service with the U.S. Army [29] and the BA609 [30] civil tilt-rotor aircraft developed jointly by Bell and Agusta helicopter company (Fig 5c) show that the tilt-rotor aircraft is the most successful new type of high-speed rotorcraft so far. In addition, the tilt-multi-rotor aircraft with distributed electric propulsion (DEP) system is also relatively rapidly developed, among which the representative aircrafts are: Beta technologies, AVA XC [31] (Fig. 5d) of Inc., which adopts tilt-coaxial rigid rotor system to ensure rotor aerodynamic efficiency under the premise of the body size limitation, has entered flight test stage; Joby S4 [10] (Fig. 5e) of Joby aviation adopts multi-rotor symmetrical design and has entered flight test certification stage. In addition, Uber [6] has proposed the concept aircraft ecrm-001 (Fig. 5f), which refers to the idea of compound rotor aircraft, that is, the rotors on both sides of the wing are set to tilting mode, and the remaining rotors are responsible for providing lift.

IOP Conf. Series: Materials Science and Engineering 1102 (2021) 012006 doi:10.1088/1757-899X/1102/1/012006

In addition, due to the development of DEP program, the multi-rotor can achieve stepwise tilting. That is, part of the rotors tilt into the fixed-wing mode to increase the flight speed, and the others maintain the helicopter mode to keep the lift. For different flight tasks, there can also be a variety of stepwise tilting strategies to complete more efficient flight tasks, and even be achieved in a hovering state.

(a) XV-15

(b) V-22 Osprey

(c) BA609

(d) Ava XC

(e) Joby S4

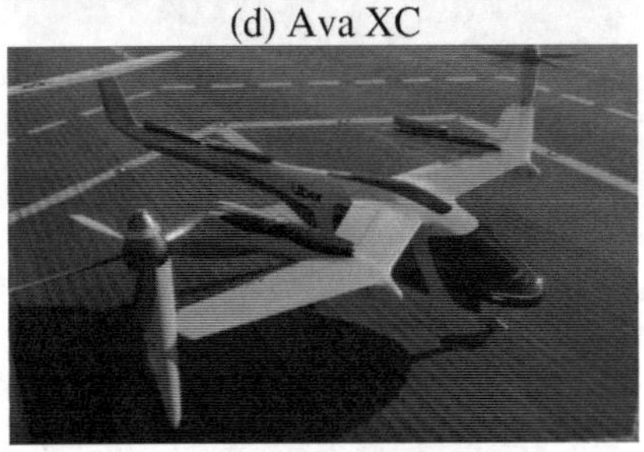
(f) eCRM-001

Figure 5. Tilt-rotor aircraft

2.3.2. Tilt-ducted fan aircraft. In order to further reduce noise and improve safety, many companies have proposed tilt-ducted fan aircraft, among which the representative models (or scaled prototypes) are: Aurora Flight Sciences XV-24A (flight test of scaled prototype [32], Figure 6a) and Lilium Jet of Lilium GmbH (flight test [33], Figure 6b). These two projects integrate multiple small ducted fans on the wing for tilting (DEP configuration), which provides plenty of propulsion redundancy under limited scale. So that safe landing can still be guaranteed in the case of partial power failure. In addition, there are more projects, like Moller International's Skycar M400 (Figure 6c, in the flight test phase [34]), XTI Aircraft Company's TriFan 600 (Figure 6d, in the flight test phase [35]), AgustaWestland's Project Zero (Figure 6e, in the flight test phase [36]) and Bell Helicopter's Nexus 6HX (Figure 6f, prototype display [37]).

Due to the existence of duct, the tilt-ducted fan aircraft can effectively reduce rotor noise. However, according to Figure 1, it can be seen that the weight of the airframe structure will increase because of the duct, and the rotor size is limited, thus the hover efficiency will be decreased.

(a) XV-24A (b) Lilium Jet

(c) Skycar M400 (d) TriFan 600

(e) Project Zero (f) Bell Nexus 6HX

Figure 6. Tilt-ducted fan aircraft

2.3.3. Tilt-wing aircraft. The principle of tilt-wing aircraft is to change the direction of the rotor thrust by tilting the wing of aircraft to achieve the functions of VTOL and high-speed cruise [38]. Compared with the tilt-rotor, here the wing and the rotors tilt together, so in the VTOL mode, the wing does not have the downward load from rotors, and the payload of the aircraft can be increased. In general, the rotor configuration of the tilt-wing aircraft is closer to the fixed-wing propeller, suitable for axial flow. Therefore, the tilt-wing aircraft can fly faster than the tilt-rotor aircraft in the high-speed cruise mode. However, due to the smaller radius of the blades and the larger load of the rotor disc, the hovering efficiency is lower. In addition, since all the rotor components rely on the wing for tilting, the tilt-wing aircraft cannot perform stepwise tilting, nor can it achieve the tilting process at low speed.

The tilt-wing technology started with the VZ-2 tilt-wing prototype in 1950s, and then the improved XC-142 tilt-wing prototype (Figure 7a) [39] successfully carried out VTOL on an aircraft carrier. The CL-84 tilt-wing aircraft built in the 1960s ~ 1970s has successfully conducted hundreds of flight tests for search, rescue, transportation, and surveillance missions [40]. In recent years, the tilt-wing aircraft using DEP has also been rapidly developed. Among them, the representative aircrafts are: Airbus's A^3 Vahana (Figure 7b, adopting eight-rotor DEP setup, in flight test stage [14]), Opener, Inc.'s BlackFly (Figure 7c, in flight test stage [41]), NASA Langley's Greased Lightning [42] (Figure 7d, scaled prototype flight test stage), and ASX MOBi from Airspace Experience Technologies (Figure 7e, scaled prototype flight test [43]), Flying Gondola of Japan Aeronext Inc. (Figure 7f, scaled prototype flight test [44]), PteroDynamics, Inc.'s Transwing (Figure 7g, tilt-wing plus fold-wing design, to save space when hovering, prototype flight test stage [45]), Bell Helicopter's Autonomous Pod Transport (APT) 70 (Figure 7h, UAV [46], the whole aircraft tilts with wings).

(a) XC-142 (b) A³ Vahana
(c) BlackFly (d) Greased Lightning (GL-10)
(e) ASX MOBi (f) Flying Gondola
(g) Transwing (h) Bell APT70

Figure 7. Tilt-wing aircraft

Studies have shown that the structure weight of the tilting mechanism can be increased significantly by using both of variable-pitch rotors/propellers and tilt-wing, especially when the wing is equipped with more than 8 rotor components. Therefore, in order to develop further, tilt-wing aircraft must overcome the above challenges in the future.

2.4. Development of advanced rotorcraft in the further

Based on the research, the current development of mature advanced rotorcraft technology can be divided into the following aspects: the compound technology of mixing the wing and the multi-rotor propulsion system, the tilting rotor technology of changing the direction of the rotor thrust to adapt VTOL and high-speed flight, and distributed electric propulsion (DEP) technology for energy and propulsion redundancy [47~49]. With the development of DEP technology, the propulsion redundancy is effectively improved due to the quantity of rotors. When a certain number of rotors or propellers fail, the remaining rotors or propellers can still ensure a safe landing of the aircraft, or even continue to complete flight tasks. Therefore, in recent years, DEP technology has been widely used in the design of new advanced rotorcraft.

The compound high-speed rotorcraft combined with DEP technology can form various advantages. The aerodynamic direction of lift system and thrust system are basically perpendicular to each other, thus the coupling effect is low. Therefore, the control strategy is relatively simple, and the effect of propulsion redundancy of DEP is better. However, its disadvantages are also prominent. This technology requires multiple rotors providing vertical lift and propulsion propellers providing horizontal thrust. Thus, the propulsion system and its auxiliary structures have a high weight ratio, especially, the horizontal propulsion system not participating in the VTOL phase results in more waste of structural weight. In addition, during forward flight at high speed, the rotation velocity of the lift rotor is reduced, thus the reverse flow region gets large. Also, the aerodynamic environment of the rotors gets harsh, which affects the dynamic characteristics of the rotors, the handling quality and the stability of the airframe. Therefore, the improvement of forward speed is also greatly restricted.

The combination of tilt-rotor aircraft and DEP technology forms the tilt-multi-rotor aircraft, which can effectively improve the performance of rotorcraft. Due to the negative twist angle of tilt rotor blade is larger, its hovering performance is lower than that of conventional helicopter. However, the hovering performance of tilt-rotor aircraft can be similar to that of the compound rotorcraft, or even better, owing to all the power systems participating in the VTOL procedure. On the other hand, compared with the compound rotorcraft, all the propulsion systems of the tilt-rotor aircraft can provide horizontal thrust at high speed, and the rotors will not encounter the reverse flow region, which makes the aerodynamic environment more stable and higher flight speed be achieved. In addition, due to the application of DEP technology, multiple rotors can tilt stepwise. That is, some rotors tilt to fixed wing mode to improve flight speed, and the other parts remain in helicopter mode to maintain lift. At different flight speeds, there can make various stepwise tilting schemes to complete flight missions more efficiently.

Therefore, this paper believes that the tilt-multi-rotor aircraft based on DEP technology meets the performance requirements of future VTOL aircraft, especially the urban aircraft. Its characteristics like efficient hovering performance, high-speed cruise capability, long range, long endurance, and high safety, environmental benign, low noise and so on, will be a feasible development direction of the new configuration rotorcraft in the future.

2.5. Key technologies of tilt-multi-rotor aircraft

Tilt-multi-rotor aircraft has obvious advantages in aerodynamic layout and propulsion system, but there are also technical challenges as follows: 1. The tilting process involves changes in the configuration and flight speed, which not only causes complicated unsteady aerodynamic effects on rotors, wings and other components [50], but also causes nonlinear variation in parameters such as the center of mass and inertia of the aircraft, resulting in additional inertial effects [51]. Therefore, the conversion control strategy between helicopter control mode and fixed wing aircraft is also complex [52], especially for a tilt-multi-rotor aircraft. 2. Additional actuating device and auxiliary structure weight are needed for tilting mechanism; 3. When DEP technology is used, the above two key technologies will be more complicated. This is because multi-rotor of DEP system is able to conduct stepwise tilting. Under different flight missions, there can also be various tilting solutions to execute flight missions more efficiently, which makes the design of tilting mechanism more complicated. How to design a reasonable dynamic conversion control strategy, complete the conversion of the two control modes, and design a safe and reliable tilting mechanism under the requirements of lightweight structure of the aircraft are the key technologies of tilt-multi-rotor aircraft.

3. Conclusion

This paper sorts and analyses the current advanced VTOL (Vertical Take-Off and Landing) rotorcraft with new configuration, and attempts to point out the feasible research direction in the future. Research shows that the aircraft with low disk load has good hovering efficiency, but the flight speed is difficult to further improve, however, with the increase of disk load, the hovering efficiency gradually decreases, but the maximum flight speed increases. In order to resolve this contradiction, researchers have proposed various advanced solutions. After decades of development, demonstration, and experiments, the current advanced high-speed rotorcraft mainly adopts three technical solutions: compound high-speed rotorcraft, rotor/wing conversion aircraft and tilt-rotor /wing aircraft.

The compound high-speed rotorcraft adopts mutually independent lift system and thrust system. Its thrust system can further increase the forward speed, and the control strategy is relatively simple. However, the thrust system cannot participate in hovering, thus the weight of the propulsion system is relatively high, and this solution cannot completely solve the problem of the reverse flow region of the rotor during high-speed flight.

Rotor/wing conversion aircraft turns the rotor into a wing by stopping the rotor, and completely converts from helicopter mode into fixed-wing aircraft mode, which solves the problem of conventional rotorcraft flight speed limitation. However, the key technologies involved are not mature, such as airfoil aerodynamics, blade control, rotor drive system and rotor/wing conversion. Obviously, it is only in the preliminary stage of development.

The tilt-rotor/wing aircraft can switch between VTOL mode and high-speed cruise mode through the tilting of the rotor group, which overcomes the problem of asymmetric airflow on both sides of the rotor during high-speed flight. It is perhaps the most ideal scheme at present. The key technical challenges are the complicated aerodynamic interference, the distribution of control strategy and the design of tilting mechanism.

In addition, distributed electric propulsion (DEP) technology, which considers the safety of energy and propulsion redundancy, has been widely used in the design of new advanced configurations of rotorcraft in recent years, so as to ensure that when 1 or 2 rotors (or propellers) fail, the remaining rotors (or propellers) can still guarantee a safe landing process for the aircraft, and even continue to complete specific flight tasks.

Through a detailed comparison and analysis of the research status, this paper draws a conclusion that the tilt-rotor aircraft combined with DEP technology is a feasible development direction for the new configuration of the advanced rotorcraft in the future. First of all, compared with the compound aircraft solution, all the power systems of the tilt-rotor aircraft can participate in VTOL and high-speed cruise, and the rotor will not encounter the reverse flow region during high-speed cruise, the aerodynamic environment is relatively stable, so it can achieve higher forward speed. Secondly, compared with the rotor/wing conversion aircraft solution, the key technologies involved are more mature and will be easier to implement in the next few years. Finally, compared with the tilt-wing aircraft solution, the multi-rotor system with DEP of tilt-rotor aircraft can realize stepwise tilting, that is, some rotors tilt to fixed wing mode to improve flight speed, while the other parts stay in helicopter mode to maintain lift, so as to improve the safety in the process of tilting. For different flight tasks, there can also be a variety of stepwise tilting strategy to complete flight tasks more efficiently, and even in a hovering state.

Therefore, this paper believes that the tilt-multi-rotor aircraft based on DEP technology meets the performance requirements of future VTOL aircraft, especially the urban aircraft. It has characteristics of high-efficiency hovering performance, high-speed cruise capability, large range, large endurance, high safety, environmental protection, low noise, etc., which is a feasible development direction of the future advanced rotorcraft.

Acknowledgment
This work was funded by program of Zhejiang Lab No. 2018DF0ZX01 and program of The Leading Innovation and Entrepreneurship Team in Zhejiang Province, No. 2018R01006.

References
[1] Datta A, Sitaraman J, Chopra I, et al. 2006. CFD/CSD prediction of rotor vibratory loads in high-speed flight. *Journal of aircraft*, Vol. **43**, (6); pp. 1698-709.

[2] Nixon M W. 1993. *Aeroelastic response and stability of tiltrotors with elastically-coupled composite rotor blades*. Maryland: University of Maryland; pp. 10-35.

[3] Karczewski M, Eglin P. 2014. Application of the modified blade element method for propeller blade design on high-speed helicopter X3. In: *Proceedings of International Symposium on Compressor & Turbine Flow Systems Theory & Application Areas "SYMKOM"*. Łódź, Polska, October; pp. 63-4.

[4] Fegely C, Juhasz O, Xin H, et al. 2016. Flight Dynamics and Control Modeling with System Identification Validation of the Sikorsky X2 Technology™ Demonstrator. In: *Proceedings of American Helicopter Society 72nd Annual Form*. West Palm Beach, USA: AHS, May; pp. 1612-24.

[5] Elan H. 2020. Uber Elevate turns its focus to community engagement. *eVTOL*, JAN22(2020).

[6] Ellie R. 2019. Uber Works to Make Air Taxis the Future of Urban Transport. *Government Technology*, JUL25(2019).

[7] Prouty R W. 1988. *More helicopter aerodynamics*. Peoria: PJS Publications; pp. 30-65.

[8] Wen Y W, Wen Q C, GAO Z, et al. 2000. *Application and development of modern helicopters*. Beijing: Aviation Industry Press; pp. 438-43.

[9] Jenkins D R, Landis T, Miller J. 2003. American X-vehicles: an inventory, X-1 to X-50. *NASA report* SP-2003-4531; pp. 33-60.

[10] Brian G G. 2020. Uber reportedly in talks to offload Elevate to Joby Aviation. *eVTOL*, DEC2(2020).

[11] Kyla L. 2020. Santa Cruz aviation company anticipates flying cars in 3 years with Marina-based factory. *KSBW Action 8 News*, JAN17(2020).

[12] Umer M, Kazmi S M A, Askari S M H, et al. 2018. Design and Modeling of VTOL Tri Tilt-rotor Aircraft. In: *Proceedings of 15th International Conference on Smart Cities: Improving Quality of Life Using ICT & IoT (HONET-ICT)*. Halifax, Canada: IEEE; pp. 1-5.

[13] Muscarello V, Colombo F, Quaranta G, et al. 2018. Aeroelastic Rotorcraft–Pilot Couplings in Tiltrotor Aircraft. *Journal of Guidance, Control, and Dynamics*, Vol. **1**, (1); pp. 1-14.

[14] Woodrow B. 2019. Airbus A³'s Vahana Electric Air Taxi Has a Single Screen Pilotless Cockpit. *Rotor & Wing International*, MAY21(2019).

[15] Loz B. 2019. Watch Vahana's flying car transition from VTOL to forward flight mode. *New Atlas*, FEB24(2019).

[16] Yeo H, Johnson W. 2009. Optimum design of a compound helicopter. *Journal of aircraft*, Vol. **46**, (4); p.19.

[17] Floros M W, Johnson W. 2004. Performance analysis of the slowed-rotor compound helicopter configuration. In: *Proceedings of the AHS 4th Decennial Specialists' Conference on Aeromechanics*. San Francisco, USA: AHS, January; pp. 1-12.

[18] Matthew W, Floros, Johnson W. 2007. Stability and control analysis of the slowed-rotor compound helicopter configuration. *Journal of the American Helicopter Society*, Vol. **52**, (3); pp. 239-253.

[19] Geiger B, Piasecki F W, Horn J F, et al. 2010. Challenges of flight control in a compound helicopter. In: *Proceedings of International Powered Lift Conference 2010*. Philadelphia: AIAA; pp. 101-13.

[20] Megan H. 2019. Boeing Autonomous Passenger Air Vehicle Completes First Flight. *CISION PR Newswire*, JAN23(2019).

[21] Kerry L. 2019. AutoFlightX Takes Wraps Off All-Electric Demonstrator. *Aviation International News Online*, APR10(2019).

[22] Samantha M. 2019. A new generation of flying cars is taking to the air — without the cars, *Providence Journal*. MAR1(2019).

[23] Elan H. 2020. Wisk's electric air taxis will fly themselves, says CEO. *eVTOL*, JAN20(2020).

[24] Eric A. 2020. A New Air Taxi Model Takes Design Cues from a Far-Flying Bird. *Wired*, JUNE12(2020).

[25] Lyu W L, Xu G H. 2015. New-Trim-Method-Based Investigation on the Cyclic-Pitch-Effected Advancing-Blade-Concept Helicopter Aerodynamics. *Journal of Aircraft*, Vol. **52**, (4); pp. 1365-71.

[26] Ruddell A J, Macrino J A. 1980. Advancing blade concept (ABC) high speed development. In: *Proceedings of American Helicopter Society 36th Annual Forum*. Washington, DC: American Helicopter Society; pp. 1-13.

[27] Newman S. 1997. The compound helicopter configuration and the helicopter speed trap. *Aircraft Engineering and Aerospace Technology*, Vol. **69**, (5); pp. 407-13.

[28] Ferguson S W. 1989. Development and validation of a simulation for a generic tilt-proprotor aircraft. *NASA report* CR-166537; pp A1-A185.

[29] Klein G D, Roberts B, Seymour C. 2000. MV-22 handling qualities flight test summary. In: *Proceedings of American Helicopter Society Annual Forum.* Washington, DC: American Helicopter Society; p. 839.

[30] King D W, Bianco-Mengotti R, Fortenbaugh R L, et al. 2005. A systems engineering approach to carefree maneuvering in the BA609. In: *Proceedings of American Helicopter Society Annual Forum.* Washington, DC: American Helicopter Society; pp. 160-70.

[31] Eric A. 2019. Beta Technologies, a Vermont Air Taxi Start-Up, Might Be About to Change the Aviation World. *The Drive*, JAN11(2019).

[32] Giannini F, Gomez M, Cottrell D, et al. 2017. Hybrid Propulsion Vertical Take-Off and Landing Aircraft, *US Patent Application* 15403818; pp. 2-10.

[33] Jeremy B. 2019. Lilium Says Its Electric Aircraft Has Pulled Off the Key Maneuver in Its Ambitious Design. *Forbes*, OCT22(2019).

[34] Andrew J H. 2017. You can now buy the Moller Skycar, one of the world's most iconic (and dubious) 'flying cars'. *The Verge*, JULY6(2017).

[35] Diane S. 2017. XTI Aircraft Company Accepts More Orders for TriFan 600 Vertical Takeoff Airplane. *PR Newswire*, OCT18(2017).

[36] Frank C. 2019. Turning Volts to VTOL. *Vertiflite*, January/February(2018).

[37] Lindsay B. 2019. Fast 5: Bell Strives for Urban Air Mobility Adoption by 2025. *MRO Network*, APR5(2019).

[38] Lindqvist A, Fresk E, Nikolakopoulos G, et al. 2015. Optimal design and modeling of a tilt-wing aircraft. In: *Proceedings of mediterranean conference on control and automation*; pp.701-8.

[39] Konrad J. 1999. Flight and operational suitability testing of the XC-142, V/STOL assault transport - Simulation and Support Conference (AIAA). *Mathematica Scandinavica*, Vol. **84**, (1); pp. 93-118.

[40] Mcgregor D M. 1965. Simulation of the Canadair CL-84 tilt-wing aircraft using an airborne V/Stol simulator. *National Research Council, Aeronautical Report LR*; pp. 19-65.

[41] Kenneth I S. 2019. Oshkosh e-AirVenture. *Vertiflite*, SEP/OCT(2019).

[42] Cooper J R, Ackerman K A, Rothhaar P M. 2018. Autonomous Path-Following for a Tilt-Wing, Distributed Electric Propulsion, Vertical Take-Off and Landing Unmanned Aerial System in Hover Mode. *NASA report* TM-2018-220109; pp. 19-46.

[43] Addie W. 2018. Detroit's ASX Reveals Electric VTOL Aircraft at 2018 North American International Auto Show. *dbusiness*, FEB12(2018).

[44] Hubert N. 2019. Flying Gondola From Aeronext: Comfortable Drone Transport. *UberGizmo*, OCT15(2019).

[45] PteroDynamics. 2019. Examples of vehicle transitions for PteroDynamics' distance VTOL Transwing™ aircraft design. *Youtube*, MARCH4(2019).

[46] Eric T. 2020. Bell Learned That Its APT 70 Cargo Drone Can Integrate With Manned Traffic During A Demo Flight. What Did It Learn About Avoidance and 5G?. *Forbes*, OCT2(2020).

[47] Stoll A M, Bevirt J B, Moore M D, et al. 2014. Drag reduction through distributed electric propulsion. In: *Proceedings of 14th AIAA aviation technology, integration, and operations conference*; p. 2851.

[48] Borer N K, Patterson M D, Viken J K, et al. 2016. Design and performance of the NASA SCEPTOR distributed electric propulsion flight demonstrator. In: *Proceedings of 16th AIAA Aviation Technology, Integration, and Operations Conference*; p. 3920.

[49] Kim H D, Perry A T, Ansell P J. 2018. A review of distributed electric propulsion concepts for air vehicle technology. In: *Proceedings of 2018 AIAA/IEEE Electric Aircraft Technologies Symposium (EATS)*. IEEE; pp. 1-21.

[50] McCroskey W J, Spalart P. 1985. Airloads on bluff bodies, with application to the rotor-induced downloads on tilt-rotor aircraft. *Vertica*, Vol, **9**, (1); pp. 1-11.

[51] Ferguson S W. 1988. A mathematical model for real time flight simulation of a generic tilt-rotor aircraft. *NASA report* CR 166536; pp. 12-36.

[52] Righetti A, Muscarello V, Quaranta G. 2017. Linear Parameter Varying Models for the Optimization of Tiltrotor Conversion Maneuver. In: *Proceedings of AHS 73rd Annual Forum*, Fort Worth, Texas, USA: AHS; pp. 280-7.

IOP Conf. Series: Materials Science and Engineering 1102 (2021) 012007 doi:10.1088/1757-899X/1102/1/012007

Utilization of Low-Potential Energy Through to the Use in Aircraft Fuel System

Renat R. Badykov[1,a], Roman A. Panshin[2,b,*], Olga V. Tremkina[2,c] and Anastasia A. Prokofieva[2,d]

[1]Department of Construction and Design of Aircraft Engines, Samara National Research UniversitySamara, Russia;
[2]Department of Heat Engineering and Heat Engines, Samara National Research UniversitySamara, Russia

[a]renatbadykov@gmail.com; [b] ssau@ssau.ru; [c]t.olga.vit@bk.ru;
[d]anastassiaofficial@gmail.com

Abstract: The article discusses the possibility of improving the main characteristics (efficiency, specific thrust) of a gas turbine engine by using liquefied natural gas (LNG) as fuel. This possibility is considered on the example of using as fuel instead of aviation kerosene on PS-90A and NK-93 engines. The parameters of the engine were determined at various stages in the cruise flight mode when operating on kerosene and LNG. The parameters evaluating the fuel consumption of the engine are calculated and compared for both fuels. The performance characteristics for kerosene and LNG were evaluated in the system of the Tu-330 subsonic transport aircraft. Energy losses during storage of cryogenic fuel on board the aircraft are estimated.
Keywords: liquefied natural gas, gas turbine engine, power plant.

1. INTRODUCTION

The use of LNG as an alternative fuel for aviation is based on the possibilities of its production in various regions of the country with virtually no serious damage to the consumption of natural gas by industry and the public. LNG is more environmentally friendly and safe to use, and also makes it possible to provide such technical characteristics of aircraft that are not achievable with traditional petroleum fuel [1]. Cryogenic fuels have higher energy performance, have high environmental quality, andtheir use will allow you to move on to new stages in the development of the aircraft industry.

Currently, modifications are being made to the Tu-330 cargo aircraft using LNG as fuel: the Tu-330-LNG military transport aircraft and the Tu-330T-LNG tanker aircraft, in which PS-90A dual-circuit turbojet engines canbe used as a power plant NK-93.

Since a sufficient amount of energy is spent during liquefaction. In the process of cooling down the circulation-pressurization system or supply of cryogenic fuel to the combustion chamber of the engine, part of the previously spent energy can be returned. Energy return can be carried out in several ways, one of which is the use of power plants operating according to Rankine or Brighton cycles [2, 3]. The electricity generated during the operation of these auxiliary cycles can be used for the operation of on-board systems, and can also be accumulated in special storage devices for further operation.

This article discusses the possibility of using the installation of generating additional energy as part of the fuel complex of the Tu-330. It is proposed to use a heat engine operating on the Rankine cycle or on the Brayton cycle as such a plant. The installation can operate on various cryogenic substances: nitrogen, krypton, argon,air, ethane, R12 freon, R22 freon, R134a freon and others [4, 5].

The scheme of the fuel system, with the installation for generating additional energy included in it, is presented in Figures 1 and 2.

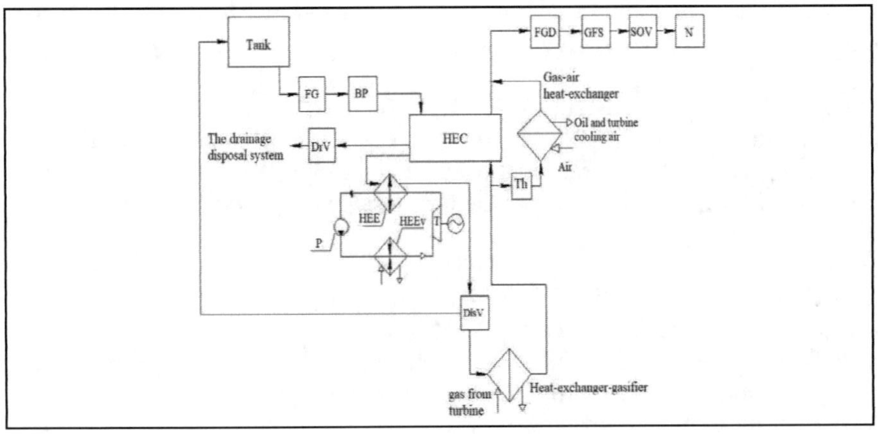

Fig. 1. The scheme of the fuel system with the installation of generating additional energy according to the Rankine cycle

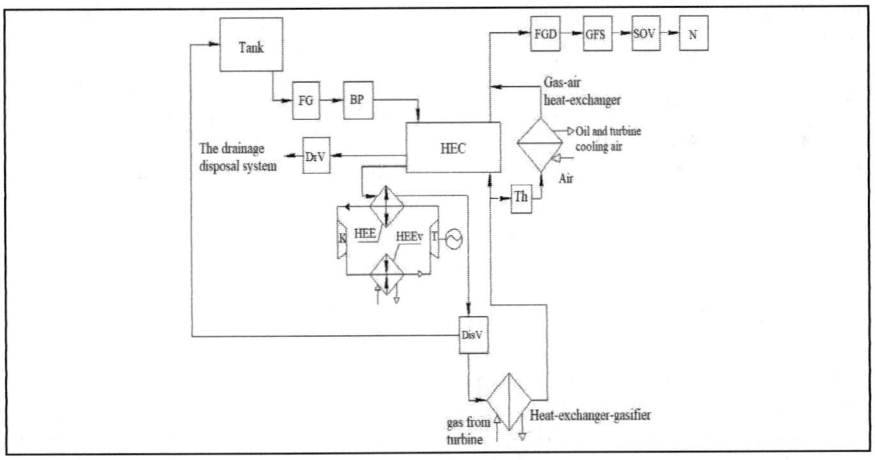

Fig. 2. The scheme of the fuel system with the installation of generating additional energy according to the Brayton cycle.

Nomenclatu

BP	booster pump;	HEEv	heat exchanger-evaporator;
DisV	distribution valve;	K	compressor;
DrV	drain valve;	N	nozzles.
FG	fire hydrant;	SOV	shutoff valve;
FGD	fuel gas dispenser;	T	turbine;
GFS	gas flow sensor;	Th	throttle;
HEC	turbopump unit;	P	Pump
HEE	heat exchanger-condenser;		

The installation is as follows. In the process of supplying cryogenic fuel to the combustion chamber of the engine or in the cooling process, when only the circuit of the circulation-pressurization system is working, the LNG passes through the heat exchanger-condenser of the additional energy generation unit. As a result, condensation of the working fluid used in the installation cycle occurs. Then there is an increase in the pressure of the working fluid (compressor or condensate pump). After increasing the pressure, the working fluid enters the heat exchanger-evaporator, and then into the turbine, where the energy of the working fluid is converted into mechanical work on the turbine shaft. The conversion of turbine energy into electricity is carried out by an electric generator.

2. CALCULATION OF THERMODYNAMIC CHARACTERISTICS

Based on well-known calculation methods [6,7], the prospects of such a solution were evaluated for various cryogenic substances suitable for operation at a temperature of 100-120 K in a liquid phase. The heat of the environment (T_H = 288 K) was used as the heat supplied to the working fluid.

For the calculation, the following initial data were used:

- at cruising flight mode, fuel consumption in the engine PS-90A G_{LNG} = 0.548 kg/s;
- at cruising flight mode, fuel consumption in theengine NK-93 G_{LNG} = 0.345 kg/s;
- fuel temperature at the inlet to the fuel assembly$T_{in.LNG}$ = 110 K.
- the temperature of the fuel at the outlet of the fuelsystem of the $T_{out.LNG}$ = 115 K.

In the Brayton cycle, the working fluid warms up toambient temperature (T_3 = 288 K).

The general expression for the thermal efficiency of thecycle is:

$$\eta_t = \frac{l_c}{q_1} = 1 - \frac{q_2}{q_1} \qquad (1)$$

where l_c – specific work of the cycle, q_1 - amount of heat supplied to the working substance, q_2 - amount of heat removed from the working substance.

The power generated by the turbine can be defined as:

$$N = l_c \cdot G \qquad (2)$$

where G – flow rate of the working substance.

An estimate of the amount of additional electrical energy received is given in tables 1 and 2.

TABLE I. ESTIMATION OF THE AMOUNT OF ADDITIONAL ELECTRIC ENERGY RECEIVED IN A UNIT OPERATING ACCORDING TO THE RANKINE CYCLE

				PS-90A (G_{LNG} = 0.548 kg/s)			NK-93 (G_{LNG} = 0.345 kg/s)		
Cryoproduct	**q_1, kJ/kg**	**q_2, kJ/kg**	**l_c, kJ/kg**	**G, kg/s**	**N, kW**	**η_t, %**	**G, kg/s**	**N, kW**	**η_t, %**
Nitrogen	90	89	1	2.6	2,6	1.1	1.6	1.6	1,1
Nitrogen	92	89	3	2.6	7,8	3.3	1.6	4.9	3.3
Krypton	113	109	3,5	2.0	7	3.1	1.3	4.4	3.1
Krypton	119	109	10	2.0	20	8.4	1.3	12.6	8.4
Krypton	123	109	14	2.0	28	11.4	1.3	17.6	11.4
Argon	130	123	7	1.8	12.4	5.4	1.1	7.8	5.4
Argon	133	123	10	1.8	17.7	7.5	1.1	11.2	7.5
Argon	140	123	17	1.8	30.1	12.1	1.1	19	12.1
Air	123	122	1	1.8	1.8	0.8	1.1	1.1	0.8
Air	125	122	3	1.8	5.4	2.4	1.1	3.4	2.4
Ethane	610	568	42	0.4	16.1	6.9	0.2	10.2	6.9
Ethane	618	568	50	0.4	19.2	8.1	0.2	12.1	11.3
Ethane	640	568	72	0.4	27.6	11.3	0.2	17.4	11.3

TABLE II. ESTIMATION OF THE AMOUNT OF ADDITIONAL ELECTRIC ENERGY

RECEIVED IN A UNIT OPERATING ACCORDING TO THE BRAYTON CYCLE

				PS-90A (G_{LNG} = 0.548 kg/s)			NK-93 (G_{LNG} = 0.345 kg/s)		
Cryoproduct	**q_1, kJ/kg**	**q_2, kJ/kg**	**l_c, kJ/kg**	**G, kg/s**	**N, kW**	**η_t, %**	**G, kg/s**	**N, kW**	**η_t, %**
Nitrogen	115	78	37	0,09	3,2	32,3	0,06	2,1	32,3
Krypton	20	12	8	0,6	4,2	41,2	0,4	2,6	41,2
Argon	40	24	17	0,3	5,4	41,3	0,2	3,4	41,3
Air	110	74	36	0,09	3,3	32,9	0,06	2,1	32,9
Ethane	96	72	24	0,08	1,8	24,7	0,05	1,2	24,7
Freon R12	18	15	3	0,2	0,6	16,4	0,1	0,4	16,4
Freon R22	14	10	3	0,4	1,4	26,7	0,2	0,9	26,7
Freon R134a	37	35	2	0,2	0,6	6,5	0,2	0,3	6,5
Nitrogen	93	56	37	0,1	4,7	39,5	0,08	3	39,5
Krypton	11	6	5	0,9	4,9	47,6	0,6	3	47,6
Argon	21	11	10	0,7	6,4	47,0	0,4	4	47,0
Air	53	88	35	0,3	5,3	39,9	0,1	3,4	39,9
Ethane	70	48	23	0,05	1,2	32,0	0,04	0,7	32,0
Freon R12	11	9	2	0,4	0,9	21,3	0,3	0,6	21,3
Freon R22	3	1	2	0,6	1,4	69,7	0,4	0,9	69,7
Freon R134a	30	29	2	0,2	0,2	5,1	0,1	0,2	5,1
Nitrogen	75	42	33	0,2	5,6	43,9	0,1	3,5	43,9
Krypton	3	2	1	3,4	4,5	42,1	2,1	2,8	42,1
Argon	5	3	2	1,1	1,2	33,1	0,7	1,1	33,1
Air	71	39	32	0,2	6,1	44,6	0,1	3,9	44,6
Ethane	51	32	20	0,2	3,1	38,2	0,1	1,9	38,2
Freon R12	6	4	2	0,7	1	25,8	0,4	0,6	25,8
Freon R134a	25	25	0,5	0,3	0,1	1,9	0,2	0,1	1,9

3. PROCESSING RESULTS

Based on the results obtained, the dependences of the amount of heat, cycle operation, plant power, thermal efficiency, substance flow rate on the degree of expansion of the gas in the turbine are obtained (Figures 3-7).

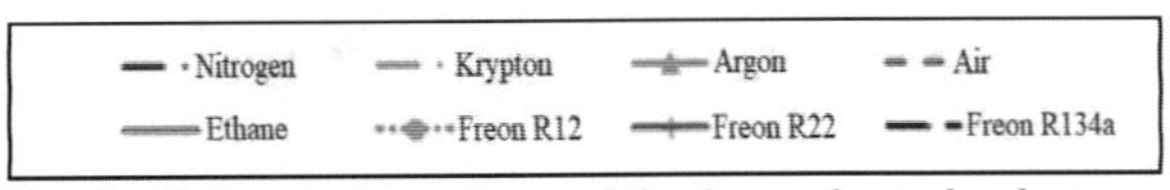

Fig. 3. Color designations of the investigated substances.

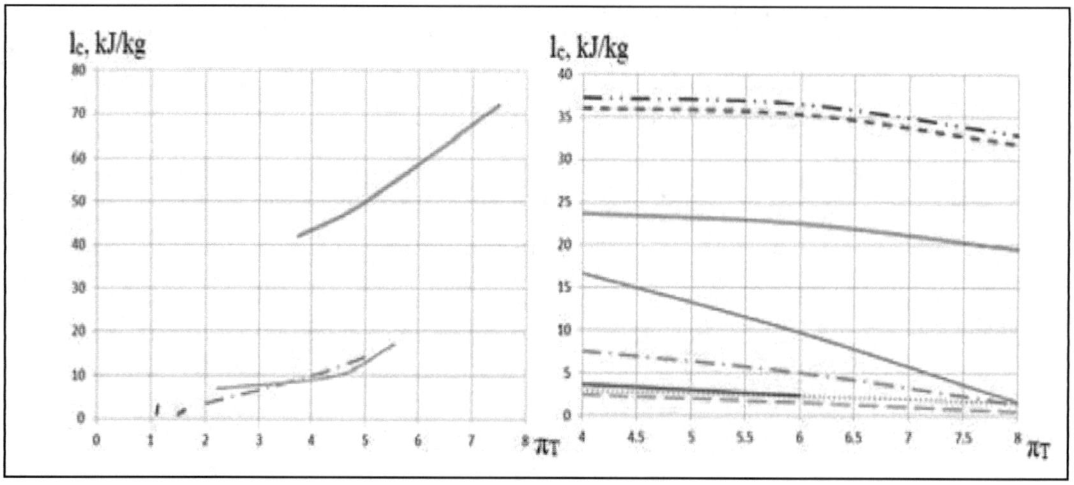

Fig. 4. The dependence of the cycle on the degree of expansion of the gas in the turbine of the unit. left) according to the Rankine cycle; right) according to the Brayton cycle.

According to the results of calculations in Figure 4, we see that in the case of using a power plant operating according to the Rankine cycle, the dependence of the cycle mode on the expansion ratio is direct in the case of operating according to the Rankine cycle. The Brayton cycle is reversed, and it should be noted that, under equal conditions, the ethane work values in the Rankine cycle are higher than in the Brayton cycle and increase with increasing expansion (compression in the cycle) in the turbine. However, the work of substances such as argon and krypton in the case of the Brayton cycle is higher than in the Rankine cycle. But as the degree of expansion increases, the value of work decreases. This is due to the fact that with a constant increase in the compression ratio in the compressor, the work losses multiply increase.

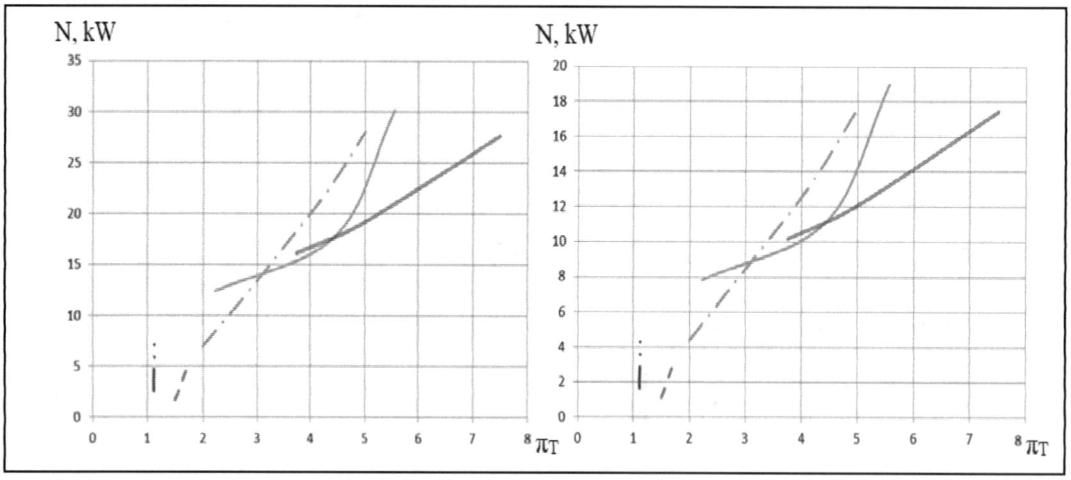

Fig. 5. Dependence of plant power on the magnitude of the degree of expansion of gas in a turbine of a plant with a Rankine cycle. left) when operating as part of the fuel circuit PS-90A (LNG); right) when operating as part of the NK-93 (LNG) fuel circuit.

In Figure 5, you can see that in the case of using the power plant of the Rankine cycle for both engines, the power lines for each substance from the expansion ratio behave the same. For example, in both cases, the dependence of power on the expansion ratio is almost linear. However, the numerical values of the generated

power in the case of the installation with the PS-90A engine are much higher than in the case of the application with the NK-93 engine.

It can also be said from the graphs that in the case of high values of the expansion ratio, it is more efficient to use argon and ethane.

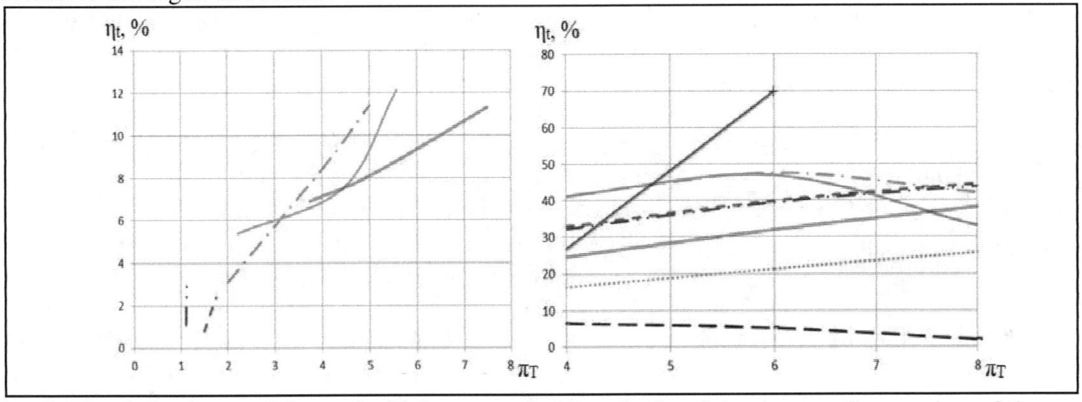

Fig. 6. The dependence of thermal efficiency on the magnitude of the degree of expansion of the gas in the turbine of the installation. left) power plants with a Rankine cycle; right) power plants with a Brayton cycle.

Considering Figure 6, we can conclude that the behavior of substances such as argon and krypton in the Rankine cycle differs from the Brayton cycle. For example, in the Rankine cycle, with an increase in the degree of expansion, the value of the thermal efficiency increases.

The power values in the case of using the working fluid of ethane in both cases change relatively the same.

It is also worth noting as the main difference between the two presented power plants that in the case of the Rankine cycle, the thermal efficiency values are significantly lower than in the application of the Brayton cycle.

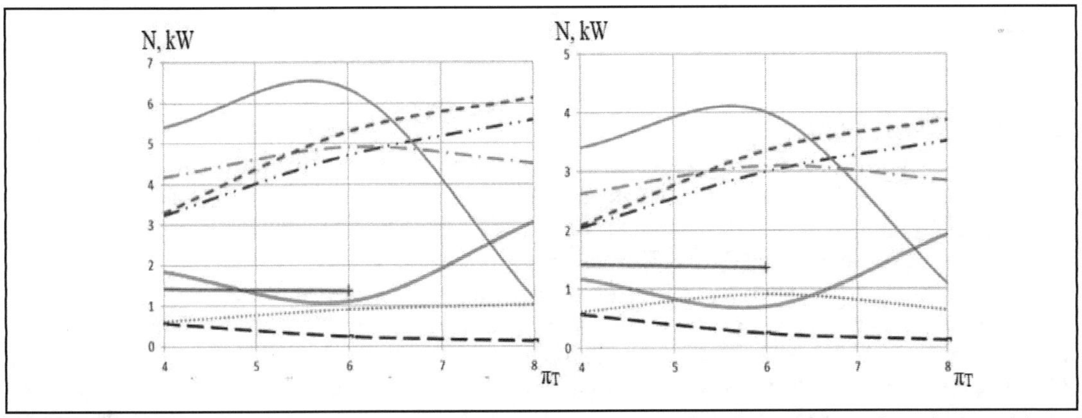

Fig. 7. Dependence of plant power on the magnitude of the degree of expansion of gas in a turbine of a plant with a Brayton cycle. left) when operating as part of the fuel circuit PS-90A (LNG); right) when operating as part of the NK-93 (LNG) fuel circuit.

Figure 7 shows that by analogy with Figure 5, the behavior of the graphs is relatively the same. And also, similar conclusions can be drawn about the values of the capabilities in the case of using each of the engines.

4. CONCLUSIONS

Thus, for the installation of generating additional energy as part of the fuel complex of the Tu-330 aircraft, the most effective is the use of power plants operating according to the Rankine cycle using argon as a working substance and a turbine with a degree of expansion of π_T=5.56. With these parameters, the power generated by the installation is the largest and amounts to N=30.141 kW and N=18.972 kW for PS-90A and NK-93 engines, respectively.

This article describes and describes the operation of the installation for generating additional energy included in the cryogenic fuel complex. The parameters of the installation based on the Rankine and Brayton cycle were calculated for various cryogenic products: nitrogen, krypton, argon, air, and ethane, R12 freon, R22 freon and R134a freon. Plots are plotted for the amount of heat input, the cycle operation, cryoproduct consumption, thermal efficiency, and the power generated by the installation, depending on the degree of expansion of the turbine. Diagrams of the Rankine and Brighton cycle are constructed for various degrees of expansion of the turbine. Analysis of the calculation results shows that it is most efficient to use power plants of the auxiliary circuit operating according to the Rankine cycle, it is advisable to use argon as the working fluid, and the degree of expansion of the turbine is π_T=5.56. Moreover, the power generated by the installation is the largest and is equal to N=30.141 kW and N=18.972 kW, when operating as part of the fuel system with PS-90A and NK-93 engines, respectively.

ACKNOWLEDGMENT

The results of the work were obtained using the equipment of the collective use center "INTER-FEDERAL TRAINING AND PRODUCTION RESEARCH CENTER OF SELF-TECHNOLOGIES" with financial support from the Ministry of Science and Higher Education of the Russian Federation (project № FSSS-2020-0019).

The results of the work were obtained using the equipment of the collective use center "EDUCATIONAL RESEARCH CENTER "VIBRATION STRENGTH AND RELIABILITY OF AEROSPACE PRODUCTS" with financial support from the Ministry of Science and Higher Education of the Russian Federation (project № FSSS- 2020-0019).

REFERENCES

[1] I. Barmin. Liquefied natural gas yesterday, today, tomorrow [Text] / I. V. Barmin, I. D. Kunis; under the editorship of A.M. Arkharov.- M.: MSTU. N.E. Bauman, 2009. -- 256 p.

[2] Song Y, Wang J, Dai Y, Zhou E. Thermodynamic analysis of a transcritical CO2 power cycle driven by solar energy with liquified natural gas as its heat sink. Appl Energy2012; 92: Pp. 194–203.

[3] Zaika A.V., Tereshchenko O.V., Uglanov D.A. et al. Determination of the effectiveness of steam turbine plants utilizing low-grade heat of a cryogenic product // XIV KOROLOVSKIY READINGS. - 2017. - T. 1. - Pp. 381-382

[4] Andreev V.A., Borisov V.D., Klimov V.T., Malyshev V.V., Orlov V.N. Caution: gases. Cryogenic fuel for aviation. - M.: Moscow Worker, 2001. -- 224 p.

[5] Vargaftik, N.B. Handbook of thermophysical properties of gases and liquids [Text] / NB. Wargaftik. - Moscow: Nauka, 1972 - 720p.

[6] Grigoryev V.A., Zhdanovsky A.V., Kuzmichev V.S., Osipov I.V., Ponomarev B.A. The choice of parameters and thermogasdynamic calculations of aircraft gas turbine engines: a training manual. - Samara: Publishing house of the Samara State Aerospace University, 2009. - 202 p.

[7] Grigoriev V.A. Design thermogasdynamic calculation of civil aviation aircraft GTE: a training manual. - Samara: Publishing House of the Samara State Aerospace University, 2001. - 170 p.

Analysis of the Impact of Airborne Multi-antenna Radar Power Architecture on Weight

Lin Zaiyu[a, *], Yu Qingqing[b] and Zhu Jun[c]

Nanjing Research Institute of Electronics Technology, Nanjing, China

[a]lzy_cet@163.com; [b]yuqingqing@aliyun.com; [c]llccjzjz@sohu.com

Abstract: Multi-antenna phased array radar is widely used in airborne detection, and the weight of its power system is affected by the number of antenna and power architecture. Based on the morphological differences of secondary power supply, the power architecture is divided into independent unit architecture, independent module architecture and distributed architecture. By analyze the weight composition and characteristics of power system in multi-antenna radar, the applicable scenarios of various architectures are obtained, and the influencing mechanism of antenna number on power weight is demonstrated. Finally, based on typical airborne fire control radar, the specific weight composition of power system of various architectures in single-antenna radar and three-antenna radar is compared, and the specific effect of antenna number on power weight is obtained.
Keywords: Multi-antenna Radar, Power Architecture, Weight

1. Introduction

Phased array radar is a detective system with multiple functions, such as rapid scanning of antenna beam, rapid changing of antenna beam shape, spatial orientation and spatial domain filtering, spatial power combining, conformal with platform and multi-beam forming. It has been widely used in space-object detection, advanced air defense systems, airborne early warning, airborne fire control, multi-target precision tracking and measurement [1].

Static single-antenna array cannot achieve omnidirectional scanning. There are two methods to realize omnidirectional beam scanning, one is the traditional mechanical scanning radar, and the other one is multi-antenna phased array radar, which arranging multiple phase arrays in different directions statically to cover all airspace. Multi-antenna phased array radar has many advantages, such as no transmission system, high reliability, strong adaptive ability and high scanning speed [2].

Different from single-antenna phase radar, the power system of multi-antenna phase radar is more complex, which needs to meet the power supply requirements of each antenna array at the same time. Based on the comprehensive consideration of the overall combat performance, the evaluation indexes of the power system are not only limited to the power supply capacity, conversion efficiency, power supply quality and other technical indexes of power supply, but also including the system weight, which is the key mechanical index of airborne equipment. Aiming at the airborne multi-antenna phase array radar, this article analyzes the weight composition and influence factors of power systems in different architectures.

2. Power system OF multi-antenna radar

a) Main form and typical power supply mode of multi-antenna phased array radar

Airborne multi-antenna phased array radar is commonly used in early warning aircraft (AEW). There are many types of AEW equipping with multi-antenna phased array radar, such as A-50 AEW (Russia), ZDK-03 AEW (China), Phalcon AEW (Israel), S100B AEW (Sweden), etc. The appearance of each AEW is shown in Fig.1. In order to expand the azimuth of sight and improve the survivability in battlefield, some types of fighter aircraft have equipped with multi-antenna phased array radar in recent years, such as Sukhoi Su-57 (Russia) shown in Fig.2.

Figure 1. Typical AEW equipping with multi-antenna radar

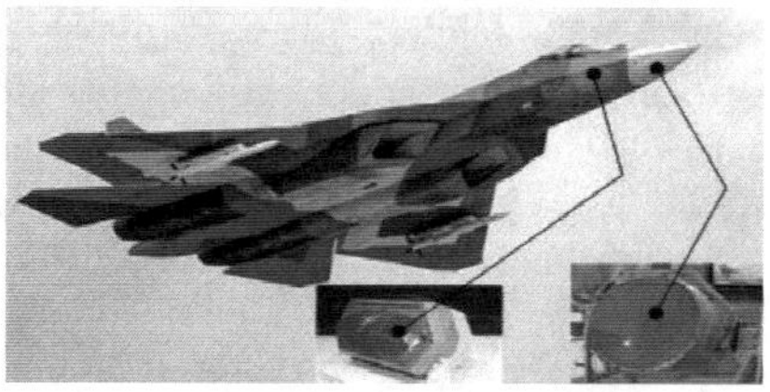

Figure 2. Typical fighter aircraft equipping with multi-antenna radar

The object of this article is the multi-antenna phase array radar with a moderate distance between each antenna and a common power system, excluding the situation that the antennas are too discrete and completely independent.

The power consumption of the antenna array unit is not always divided equally, and needs to be adjusted according to the operational requirements of the mission system in different spatial domain. Especially in the single target tracking mode, the radar needs to concentrate power supply on a certain antenna array unit. Therefore, the power system must have the ability to meet the full power supply of all antenna array units.

b) Composition of power system

Phased array radar uses transceiver modules widely, which requires a power supply with high current, high reliability and high stability characteristics. The power system generally adopts a three-level conversion architecture (shown in Fig.3): the primary power component converts the AC power provided by the aircraft into a DC high-voltage power; the secondary power component converts the DC high-voltage power into various types of DC low-voltage power; the tertiary power component converts the DC low-voltage power into various voltages for specific RF module. The tertiary power component is installed in the transceiver and RF module, and will not be analyzed in this article.

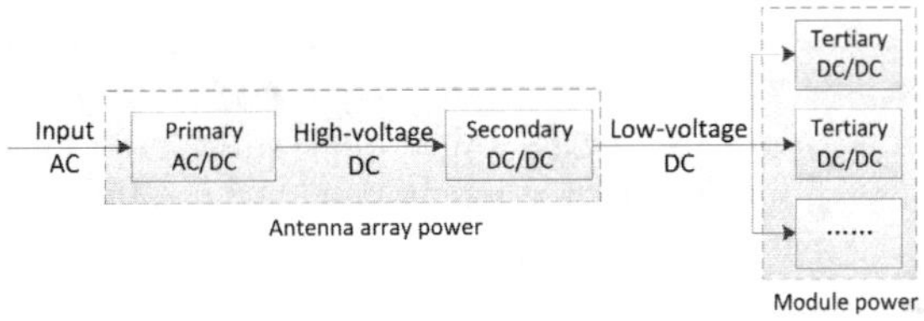

Figure 3. Three-level power conversion architecture.

c) Influencing factors of power system weight

As a relatively independent sub-system in the radar system, the weight of power system is related to the demand parameter, and is positively related to the total power, the number of power types, the quality of the output power, and the severity of the environmental conditions [3]. The influence of the power distribution architecture on the weight is presented as a non-deterministic coupling correlation, which needs to be comprehensively evaluated according to the specific radar design scheme.

3. Impact of architecture on power system weight

a) Classification of power system architecture

The differences of power system architectures analyzed in this article are mainly reflected as the differences of the form of secondary power component. According to this, the power system architecture can be divided into three types: independent unit architecture, independent module architecture, and distributed architecture (shown in Fig.4).Independent unit architecture combines the primary power component and the secondary power component into an independence unit, which is independent with antenna array unit; independent module architecture designs the primary power component as a unit which is independent with antenna array unit, and designs the secondary power component as an independent module of antenna array unit; distributed architecture designs the primary power component as a unit which is independent with antenna array unit, and designs the secondary power component as distributed small modules of antenna array unit.

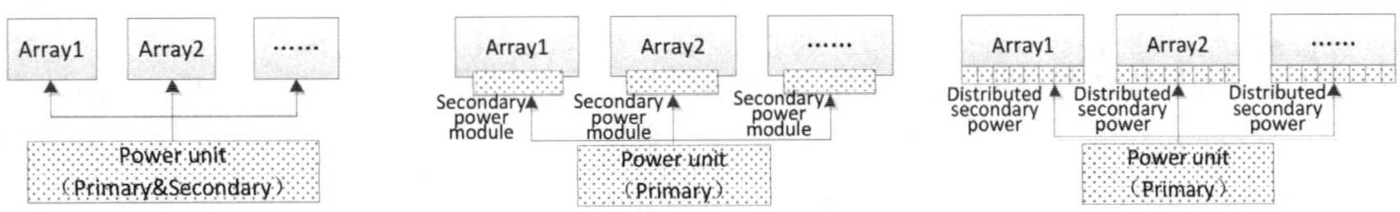

1) Independent unit 2) Independent module 3) Distributed

Figure 4. Three types of system architecture

b) System weight analysis of independent unit architecture

In this architecture, the antenna array power supply of the whole system is combined into one physical unit, including a primary power module and multiple varieties of secondary power modules. Considering the modular design and improving the maintainability, the independent power unit mostly adopts the form of "chassis-blind plug module"(shown in Fig.5), resulting in more and heavier structural parts of the power supply unit. However, the antenna array unit is light because it does not need to install secondary power component.

The secondary power component is lighter. Due to the centralized layout of the secondary power supply, the entire system only needs to equip one set of full-power secondary power devices to meet the full power supply needs of all antenna array units. Therefore, it can reduce the redundancy of the secondary power devices effectively.

The interconnecting cables are relatively heavier. First, after primary and secondary conversions, the interconnecting cables between the power unit and antenna array unit are low-voltage high-current cables, so that the weight per unit length of the cable is heavier. Second, there are more than one type of power supply cables after secondary conversion, so that the number of cables has doubled.

Base on a single perspective of system weight reduction, this architecture is suitable for radars with more antenna arrays and a shorter distance between the power unit and the antenna array unit.

c) System weight analysis of independent module architecture

In this architecture, the power system consists of one primary power unit and several secondary power modules (shown in Fig 5). Since the primary power unit has a more single function and a smaller volume, a box structure is generally used. There are no components such as the chassis and interconnecting backplane, so that the weight of primary power is lighter.

The secondary power component is heavier. First, in order to improve the maintainability, the secondary power module is mostly connected with the antenna array unit by blind plugging. Therefore, it is necessary to add structures such as guide rails and blind plug connectors. Second, each antenna array unit needs a full-power secondary power module, which causes the secondary power to double.

The interconnecting cables are relatively lighter. After primary conversion, the interconnecting cables between the power unit and antenna array unit are high-voltage low-current cables, so that the weight per unit length of the cable is lighter. Furthermore, there is only one pair of cables usually.

Base on a single perspective of system weight reduction, this architecture is suitable for radars with less antenna arrays and a longer distance between the power unit and the antenna array unit.

d) System weight analysis of distributed architecture

In this architecture, the secondary power is changed to several distributed parallel module based on the independent module architecture. This architecture has the advantages of simple thermal design, easy modularization and standardization, high redundancy and reliability, high accuracy of load voltage regulation, and easy expansion [4-6].

In this architecture, the state of the independent power unit (primary power) and the interconnecting cables are consistent with the independent module architecture. In AEW radars operating in lower frequency bands, due to the higher power and electromagnetic compatibility requirements, the secondary power devices are larger in size, and generally installed in the surrounding area of the functional module with independent structural shells or incorporated into the functional module without independent structural shells. In fighter aircraft radars operating in higher frequency bands, due to the high degree of system integration, secondary power devices can be stacked and installed into the overall structure of the antenna array unit as a profile, and a sandwich structure needs to be added.

Base on a single perspective of system weight reduction, the applicable situation of this architecture is the same as that of the independent module, but the weight of the two architectures needs to be evaluated according to the specific situation.

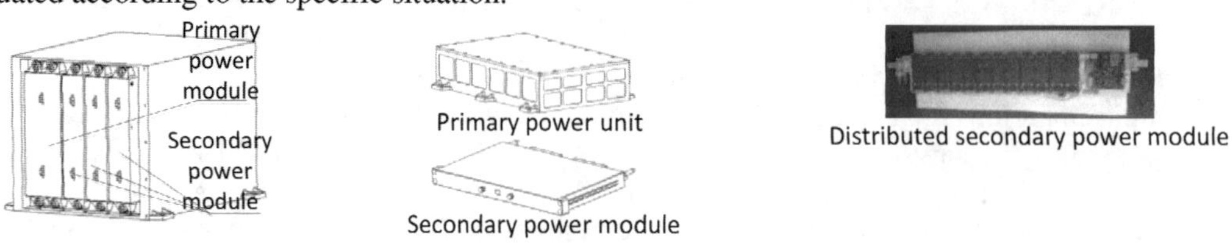

1) Independent unit 2) Independent module 3) Distributed [7]

Figure 5. The typical form of power component in different architecture

e) Impact of antenna array quantity on power system weight

The number of antenna array affects the weight composition of different power architectures directly. Under the same total power consumption conditions, when the number of antenna array is small, the independent unit architecture power system is heavier because of heavier unit chassis and interconnection cables; when the number of antenna array is large, the remaining two architectures power systems are heavier because of the doubling of secondary power devices.

4. Example analysis

a) Analysis Overview

In the actual engineering design, the choice of power architecture needs to conduct overall evaluation from multiple perspectives, such as electrical performance, structural performance, technological feasibility, quality characteristics, and cost. This example is based on some typical fighter aircraft fire control radar. Under the same total power consumption, the power system weight composition under different antenna array quantity is analyzed.

In the analysis, the weight of the power system is divided into six parts: the primary power device (recorded as PPD), the secondary power device (recorded as SPD), the primary power structure (recorded as PPS), the secondary power structure (recorded as SPS), the antenna array unit frame weight gain (recorded as AFG), and the interconnection cable (recorded as IC).

In the independent unit architecture, the primary power and the secondary power are combined into a chassis unit. The weight of the structural components of the entire chassis is divided into the PPS and SPS in proportion to the ratio between the PPD and SPD. In the independent module architecture, the SPS refers to the blind plug module structure of the secondary power, and the AFG refers to subsidiary structure added to support the installation of the secondary power module in antenna array unit, such as the directional guides, pins and blind plug connectors. In the distributed architecture, the SPS refers to the sandwich structure required for distributed installation, and the AFG refers to the mounting blocks and rigidity strengthening structures required to match the power sandwich structure.

b) Analysis example simplification

Simplify the analysis example as follows：

- Does not consider the situation that the arrays are too discrete and completely independent.
- Assume that the dimensions of each array are the same, and all arrays have full power mode.
- Assume that the total input power of the system is constant, so that the PPD is constant in various architectures and various antenna array quantities.
- Set the PPD to one unit weight, and the weight of each part is measured by the ratio of its actual weight to the PPD.

c) Power weight composition of single-antenna array radar

The power weight composition of three architectures in single-antenna array radar is shown in Fig.6. The weights of three architectures are roughly the same. The distributed architecture is the lightest and the independent unit architecture is the heaviest.

Fig.7 is a pie chart of weight composition. It can be seen that in the independent unit architecture, the chassis weight (PPS + SPS) and IC are the components that make it heaviest; In the independent module architecture, the secondary power is installed on the antenna as a module, without the need for a separate chassis and a sandwich structure, so that the SPS is reduced significantly; In distributed architecture, although the SPS in the form of sandwich has increased significantly, but the SPD has also decreased significantly.

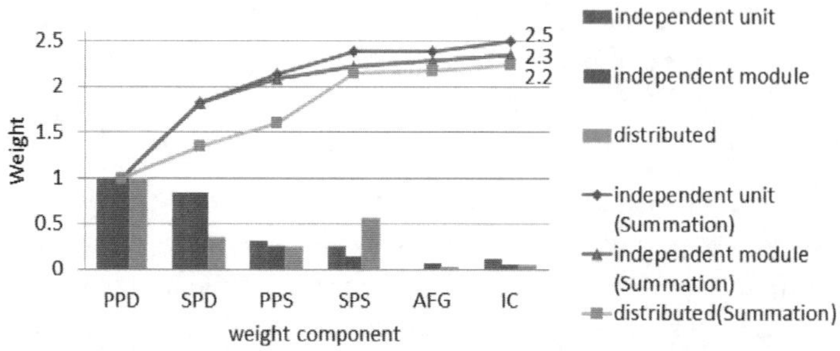

Figure 6. Power weight composition of single-antenna array radar

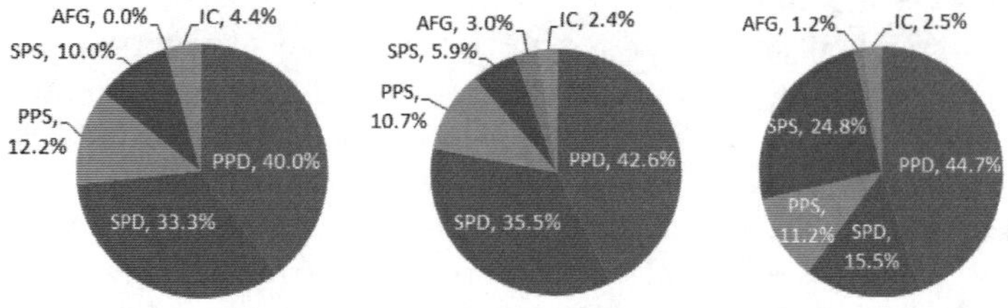

1) Independent unit 2) Independent module 3) Distributed (7)

Figure 7. Weight composition pie chart of single-antenna array radar

d) Power weight composition of three-antenna array radar

The power weight composition of three architectures in three-antenna array radar is shown in Fig.8. The weights of three architectures vary widely. The independent unit architecture is the lightest and the independent module architecture is the heaviest.

Fig.9 is a pie chart of weight composition. It can be seen that in the independent unit architecture, the total weight does not increase much compared to the single-antenna array radar, which is mainly reflected in the increase of the IC; in the other two architectures, the SPD and SPS have doubled, resulting in significant weight gain.

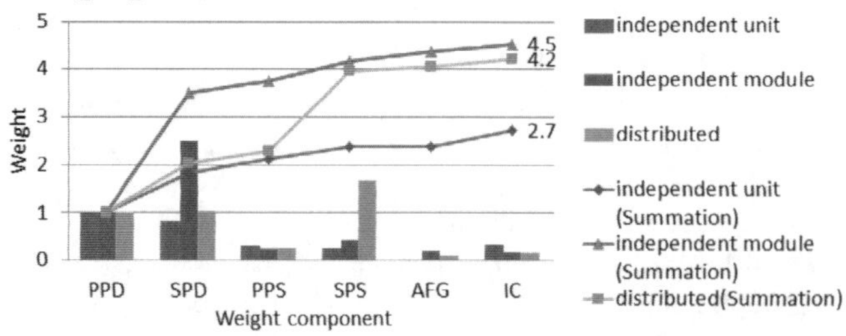

Figure 8. Power weight composition of three-antenna array radar

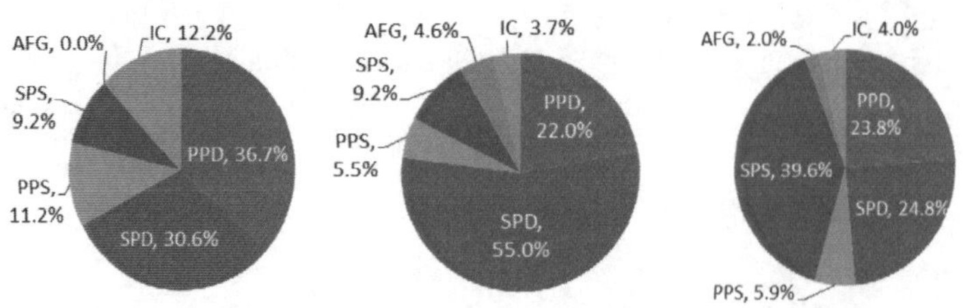

1) Independent unit 2) Independent module 3) Distributed (7)

Figure 9. Weight composition pie chart of three-antenna array radar

e) Correlation between the array quantity and the power weight

After analyzing the situation of the two-antenna array and the four-antenna array similarly, the total weight of the power system under different numbers of arrays is summarized and compared in Fig.10.

When the number of antenna arrays is small, the power weights of three architectures are close, and the distributed architecture has a slight advantage; as the number increases, the power weight of the independent module architecture and the distributed architecture increases significantly. Due to higher device integration, the weight of the distributed architecture is slightly better than the independent module architecture.

In the independent unit architecture, the secondary power supply device is shared by multiple arrays, so the weight is not sensitive to changes of array quantity. In the other two architectures, each antenna array is equipped with a secondary power separately, so the weight is sensitive to changes of array quantity.

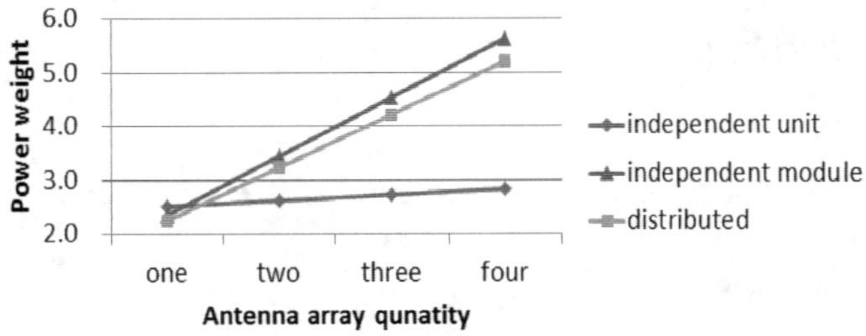

Figure 10. Comparison chart of power weight under different antenna quantity

5. Conclusion

Based on the analysis and example comparison of the weight composition of different power architecture in the multi-antenna array radar, the following conclusions are formed:

1) The antenna array quantity has different effects on the power weight of different architecture. In independent unit architecture, power weight is not sensitive to the change of array quantity; in the independent module architecture and distributed architecture, power weight is sensitive to changes of array quantity;

2) When the antenna array quantity is large, the independent unit architecture is lighter; when the antenna array quantity is small, the other two architectures are lighter;

3) The independent module architecture and the distributed architecture are very close in weight, but the distributed architecture with a higher integration has a slight advantage.

This article demonstrates the impact of the antenna array qunatity and power architecture on the weight of the radar power system, and provides a theoretical reference for the overall demonstration and design of the power system of multi-antenna phase array radar.

References

[1] Zhang Yiguang, Technical characteristics and key technologies of phased array radar[J]. Electronic Science & Technology Review,1996,(7)

[2] Lu Jun, Introduction to airborne early warning system(verision 2)[M]. National Defense Industry Press, 2011

[3] Yu Qingqing, Wang Changwu, Sensitivity analysis of system weight for radome system of AEW radar[J]. Modern Radar,2015,(6)

[4] Tabisz W A , Jovanaic M M, Lee F C. Present and Future of Distributed Power System. APEC,Blacksburg, VA, USA, IEEE,1992:11-18

[5] Cho B H, Choi B. Analysis and Design of Multi-Stage Distributed Power Systems.INTELEC, Kyoto,Japan,IEEE,1992:220-226

[6] Hua G C, Tabisz W A. Development of a DC Distributed Power System. APEC,Blacksburg,VA,USA,IEEE,1994:763-769

[7] Xu Juan, Design and realization of a distributed power system for an active phase array radar[D]. Shanghai Jiao Tong University,2010

Analysis of Sound Insulation Mechanism and Optimization Design of an Acoustic Coating Layer with Multi-section Cylindrical Cavities

J L Zhou[1,*], X J Meng[2], H D Huang Fu[3], J H Zhang[4], K Su[5], W Tao[6] and H Y Chen[7]

Northwest Institute of Nuclear Technology, Xi'an, 710024, China.

dfyx1991zjl@163.com[1,*] flyingwood@sohu.com[2] huangfuhuidong@163.com[3] dehuazjh@163.com[4] blademasterqq@163.com[5] taoweiems@163.com[6] keys6662021@163.com[7]

Abstract. Introducing the insulation layer is an effective way to reduce sound radiation from underwater structures. Impedance mismatch properties between water and layer is closely related to sound insulation performance. Finite element method with COMSOL is used to calculate sound insulation loss in this paper. In order to investigate the insulation mechanisms of coating layer embedded with multi-section cylindrical cavities, the coating layer is approximated to a homogeneous layer with equivalent material properties. The results show that acoustic coating layer with multi-section cylindrical cavities, which causes a higher impendence mismatch has a better sound insulation performance than cylindrical cavity acoustic coating layer of the same cavity volume. The coating layer with optimized multi-section cylindrical cavities structure has obvious sound insulation advantages in the range of 200Hz to 4000Hz. Finally the theoretical analysis is verified by experiments.

1. Introduction

Sound insulation of underwater structures is generally achieved by introducing an acoustic coating layer with acoustic unit. The coating layer is used to establish an acoustic barrier between underwater target and water, reducing sound waves radiating into the water. There are various types of acoustic unit could be used to reduce the radiated noise. Ivansson S M [1]studied acoustic properties of Alberich coating with superellipsoidal cavities of mixed size. Huang[2] ,Tao[3]analysis acoustic transmission characteristics of coating layer with horizontal cylindrical cavities. Based on topology optimization method, Li[4] optimized an acoustic coating to reach a better sound absorption performance. Based on the finite element method, Panigrahi S N, Jog,etal[5],zhao[6] analysized the acoustic characteristics of the coating layer on both sides of the shell. Most of the papers mainly focus on the study of acoustic insulation characteristics of the coating layer embedded with cavities, but lack of analysis of its low frequency sound insulation characteristics and mechanism. This paper draws on the multi-scale cavity design idea, design a multi-section cylindrical cavity structure[7-9], and studies its sound insulation characteristics by finite element simulation method, and reveals its sound insulation mechanism by equivalent parameter method. On this basis, the differential evolution optimization algorithm is used to optimize the cavity structure, and the cavity structure with good low-frequcecy sound insulation effect is obtained.

2. Model

The sound insulation coating which consisted of rubber matric and cavities periodically distributed in the rubber is studied in this paper. For an infinite periodic structure, one cell can be selected for analysis. Figure 1 is the schematic diagram of insulation coating model. Figure 2 is the three-dimensional finite element model of one cell. The lattice constant of the cell is a. The coating is attached to a 10mm thick steel backing. The model is infinite in the x direction. The upper and lower ends of the model are semi-infinite waters. Sound waves are incident vertically from the upper waters. Periodic boundary conditions are set on four sides of the cell. The material parameters involved in the model are shown in Table 1.

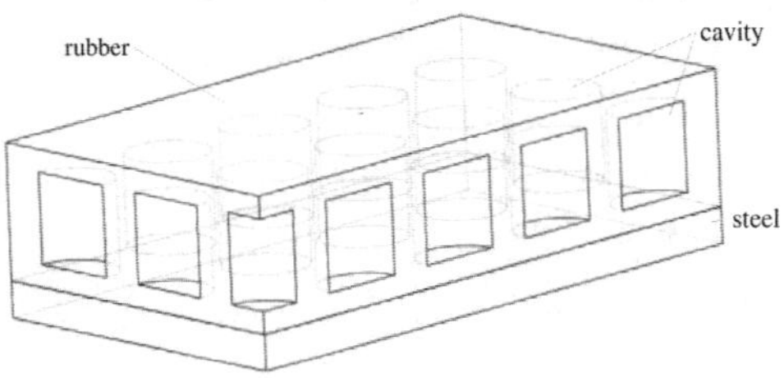

Figure 1. Schematic diagram of insulation coating model.

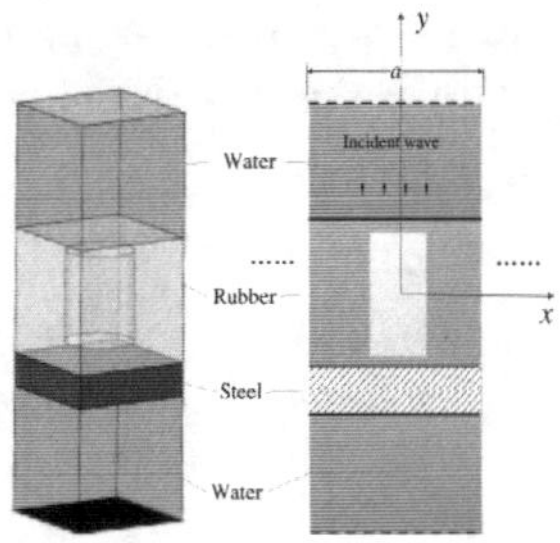

Figure 2. Three-dimensional finite element model of one cell.

Table 1. Material parameters of anechoic coating.

	Density/kg·m^{-3}	Young modulus/Pa	Poisson's ratio	Loss factor
rubber	1136.6	$3.83+1.23\times f^{0.4551}$	0.497	$0.5+0.2918\times f^{0.1471}$
steel	7850	2×10^{11}	0.33	0

f is frequency of the sound waves

3. Simulation and analysis method

In this paper, sound insulation parameter R is defined as the evaluation index of sound insulation performance, $R = I_i/I_t$, I_i represents the sound intensity of the incident waves and , I_t represents the sound intensity of transmitted waves. COMSOL Multiphysics is used for finite element modelling and analysis. In order to analyze the sound insulation mechanism of the coating containing cavities, it is equivalent to a uniform medium. The equivalent parameter of uniform material is obtained by the transfer matrix method based on acoustic parameter inversion, and the sound insulation mechanism is revealed by the change of equivalent parameters.

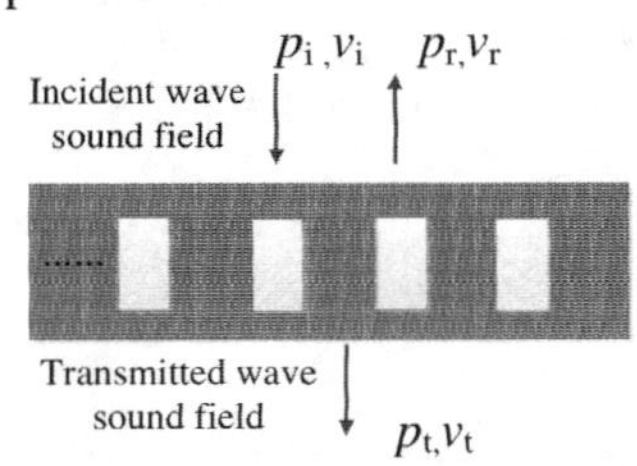

Figure 3. Plane wave transmission model of the layer.

Figure 3 shows the coating transmission model under plane wave incidence. In the incident sound field, the sound pressure and the acoustic-solid coupling surface point velocity can be expressed as:

$$p_1 = p_i + p_r \tag{1}$$

$$v_1 = v_i + v_r \tag{2}$$

In the transmitted sound field, the sound pressure and the acoustic-solid coupling surface point velocity can be expressed as:

$$p_2 = p_t \tag{3}$$

$$v_2 = v_t \tag{4}$$

As show in equation 5, the uniform coating layer has a transmission relationship between the sound pressure and the velocity of the incident surface and the transmission surface when the acoustic wave is normally incident:

$$\begin{bmatrix} p_1 \\ v_1 \end{bmatrix} = \begin{bmatrix} \cos(k_e h) & -iz_e \sin(k_e h) \\ -i\dfrac{\sin(k_e h)}{z_e} & \cos(k_e h) \end{bmatrix} \begin{bmatrix} p_2 \\ v_2 \end{bmatrix} \tag{5}$$

Suppose there is a uniform coating layer. When the sound wave is incident, the reflected sound wave and the transmitted sound wave are the same as the reflected and transmitted sound wave of the layer with cavity structure, the layer with cavities can be equivalent to a uniform coating, further obtaining the equivalent impedance z_e and equivalent wave number k_e of the coating layer.

$$z_e = -\frac{i \sin(k_e h) p_2}{v_1 - v_2 \cos(k_e h)} \tag{6}$$

$$k_e = \frac{1}{h} \arccos\left[\frac{p_2 v_2 + p_1 v_1}{p_2 v_1 + p_1 v_2}\right] \tag{7}$$

4. Sound insulation characteristics and mechanism

The acoustic coating layer exhibits good sound insulation performance because of its acoustic unit. In order to enhance its sound insulation performance and improve the low frequency sound insulation characteristics, a multi-section column formed by three cylindrical cavities with different diameters is designed. Figure 4(a) shows four different forms of the new cavity structure. The four cavities are of the same volume and same height. Sound waves are incident from the upper end of the cavity. Figure 4(b) shows the sound insulation curves of four coating layers with multi-section cavities.

(a)

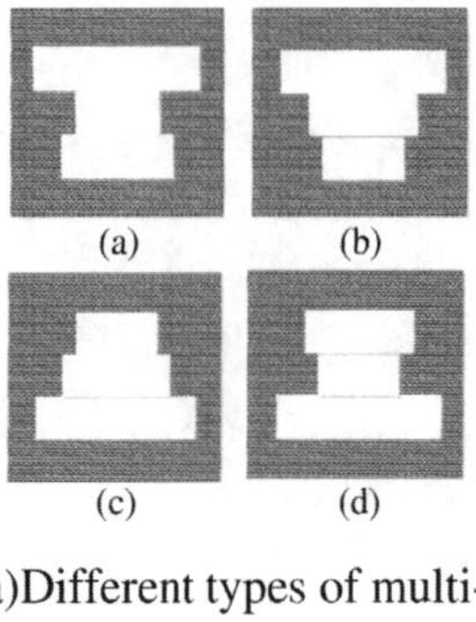

(b)

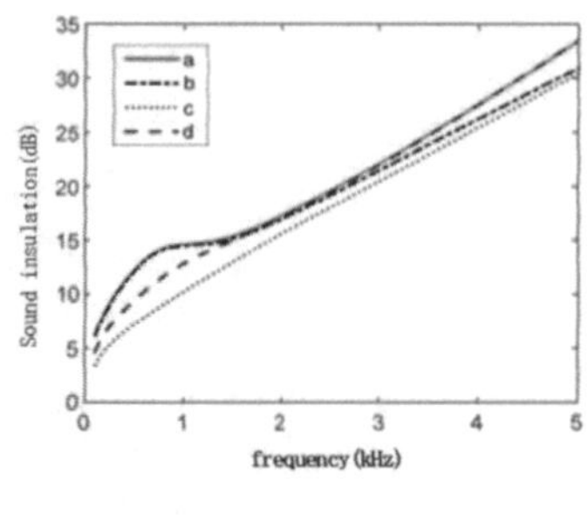

Figure 4. (a)Different types of multi-section cavity. (b)Sound insulation curve of a layer with different multi-section cavities.

It can be seen from figure 5 that when the diameter of the cavity at the incident end of the sound wave is large, the sound insulation performance in the low frequency band is good, and when the diameter of the intermediate cavity is the smallest, the sound insulation performance in the high frequency band of the cavity is good. Therefore, for a multi-section cavity structure in which three cylindrical cavities are connected in series, the two ends are large, the middle is small, and when the sound waves are incident from the large end of the diameter, the sound insulation performance is better in the entire research frequency band.

In order to reveal the sound insulation mechanism of a layer with the multi-section cylindrical cavities, the equivalent parameters of the layer is obtained by the equivalent parameter method. Define the impedance mismatch ratio as $L = |(Z_e - Z_0)/Z_0|$, Z_e is the equivalent impedance of the sound insulation coating and Z_0 is the acoustic impedance of the water. The acoustic impedance of water and uniform rubber is the same. The greater the impedance mismatch ratio, the better the sound insulation performance of the coating layer.

(a)

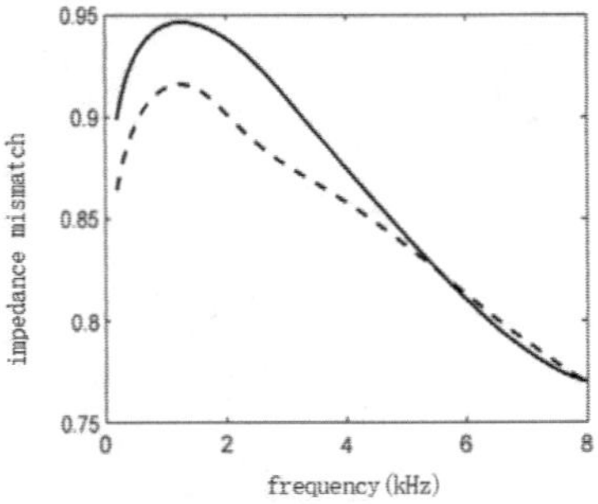

(b)

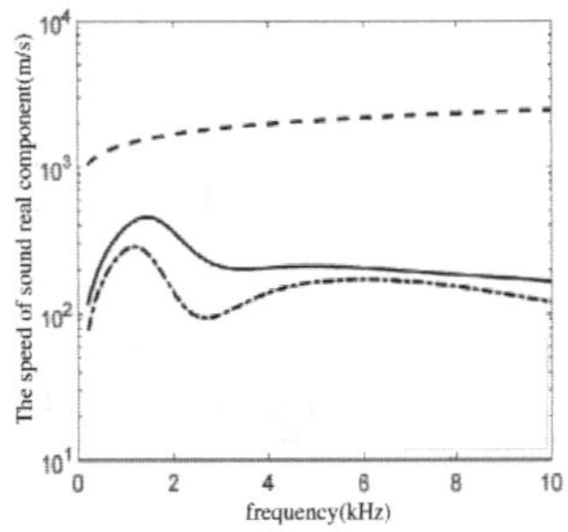

Figure 5. (a)Impedance mismatch curve.(The solid line represents the layer with multi-section cavities and the dash line represents the layer with cylindrical cavities.) (b)The real part of the velocity of sound in the layer. (The solid line represents the layer with cylindrical cavities, the dot dash line represents the layer with multi-section cavities, the dash line represents the layer of uniform material.)

The impedance mismatch ratio and the real part of equivalent sound velocity of the coating layer with multi-section cylindrical cavities and cylindrical cavities, which of the same volume and same height, are compared respectively. It can be seen from figure 5(a) that the impedance mismatch ratio of the multi-section cylindrical cavity coating layer is higher than that of the cylindrical cavity coating layer, the multi-section cylindrical cavity makes the coating layer increase the reflection of sound waves. It can be seen from figure 5(b) that the real part of the equivalent sound velocity of the matrix material is significantly reduced after the cavity is introduced, and the value of the multi-section cylindrical cavity coating layer is smaller. If the real part value of the equivalent sound velocity is small, the dissipation effect of the coating layer on the sound wave is stronger. Therefore, the multi-section cylindrical cavity coating layer improves the sound insulation performance of the coating layer by the combination of reflection and absorption.

5. Cavity structure optimization design and experimental verification

Through the above analysis, the coating layer with multi-section cylindrical cavities has better sound insulation performance than the layer with cylindrical cavity cover of the same volume. In order to obtain a multi-section cavity structure with better sound insulation effect, a differential evolution algorithm is used to optimize the cavity structure.

In this paper, the optimization goal is the transmission coefficient, and the diameter and height of three small cylindrical cavities are optimized parameters. The total volume and height of the cavity remains the same when optimized. Figure 6 shows the size diagram of the optimized model.The total volume of the cavity can be expressed as $\pi r_1^2 h_1 + \pi r_2^2 h_2 + \pi r_3^2 h_3 = \pi R^2 h_c$.,$R$ represents radius of

IOP Conf. Series: Materials Science and Engineering 1102 (2021) 012009 doi:10.1088/1757-899X/1102/1/012009

cylindrical cavity. To reduce the number of optimization parameters, define parameter $a = \sqrt{R^2 h_c/h_1}$, $b = \sqrt{R^2 h_c/h_2}$, $c = \sqrt{R^2 h_c/h_3}$, so $r_1 = a\cos\varphi$, $r_2 = b\sin\varphi\sin\theta$, $r_3 = c\sin\varphi\cos\theta$, then 6 optimization parameters can be converted into 4 optimization parameters. Table 2 shows the optimized parameters and their value ranges, and the optimal results are shown in the table3.

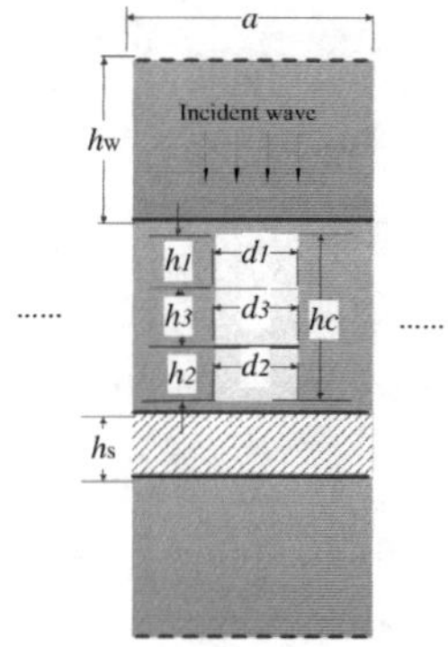

Figure 6. Size diagram of the optimized model.

Table 2. Optimization parameter table.

optimize parameters	content	value range
p_1	h_1/h_c	(0.2,0.8)
p_2	$h_2/(h_c - h_1)$	(0.2,0.8)
p_3	φ	$(\pi/6, 3\pi/8)$
p_4	θ	$(\pi/6, 3\pi/8)$

Table 3. Optimization result table.

optimize parameters	r_1	r_2	r_3	h_1	h_2	h_3
result(mm)	13.4	5.7	1.2	5.6	4.8	17.9

The optimized structure is shown in the figure 7(a). It can be seen from the results that the optimized cavity is a large two-headed structure with a small intermediate structure, and the cavity diameter of the acoustic wave incident surface is large. It can be seen from the sound insulation curve before and after optimization that the optimized multi-section cylindrical cavity has better sound insulation performance in the range of 200Hz~10kHz than the cylindrical cavity model under the same volume. The optimized structure has a sound insulation peak around 1.65 kHz, and the sound insulation advantage is obvious in the range of 200 Hz to 4 kHz.

In order to verify the reliability of the theoretical results, two experimental samples of cylindrical cavity and multi-section cavity were prepared and tested in a hydroacoustic tube. The experimental results are shown in figure 8.

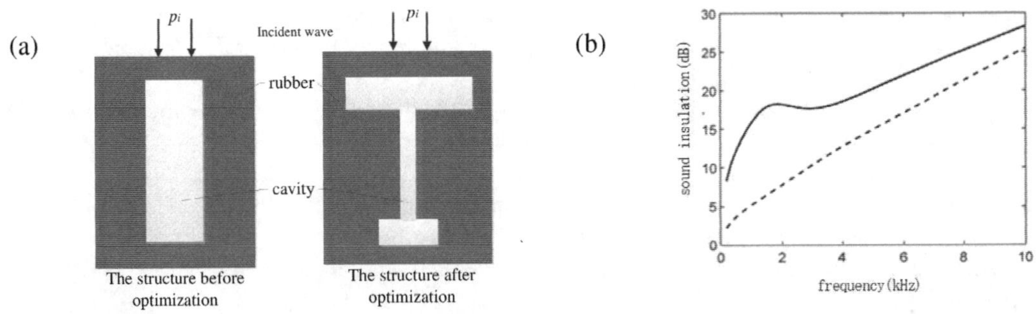

Figure 7. (a) Schematic diagram of optimized structure (b)Sound insulation curve of layer with structure before and after optimization. (The dash line represents the layer with cavities before optimization, and the solid line represents the layer with cavities after optimization.)

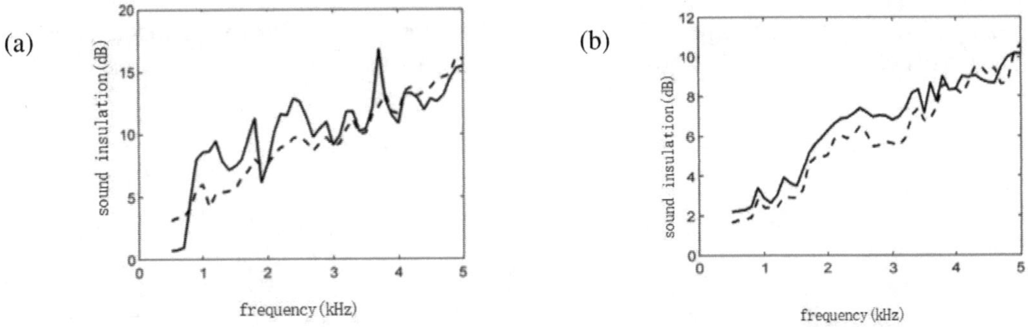

Figure 8. The sound insulation curve of layer with structure before and after optimization under 0.1MPa(a) and 1MPa(b). (The dash line represents the layer with cavities before optimization, and the solid line represents the layer with cavities after optimization.)

The sound insulation properties of the two kinds of cavity-type coverings under normal pressure and 1.0 MPa were compared. It can be seen that under normal pressure, the multi-section cavity has obvious sound insulation advantages in the low frequency range.When the pressure is increased, the sound insulation performance of the two cavity covering layers is decreased, but the sound insulation of the multi-section cavity is still higher than that of the cylindrical cavity covering layer.

6. Conclusion

Based on comsol finite element software, the sound insulation characteristics of multi-section cavity cover are analyzed. Firstly, the simplified model is verified. The equivalent parameter method is used to analyse that the multi-section cylindrical cavity cover layer has a special structure of the cavity, so that the cover layer has stronger reflection and absorption of sound waves, and the final transmitted sound wave is much lower than the same volume of the cylindrical space. Furthermore, the differential structure algorithm is used to optimize the cavity structure, and the optimal multi-section cavity structure is obtained. Finally, the results of theoretical analysis were verified by experiments.

References

[1] Ivansson S M. Numerical design of Alberich anechoic coatings with superellip soidal cavities of mixed sizeds[J].Journal of the Acoustic Society of America,2008,124(4):1974~1984.
[2] Huang ling zhi, Xiao yong, Wen jihong, Yang haibin, et al. Analysis of decoupling mechanism of an acoustic coating layer with horizontal cylindrical cavities[J].Acta Phys. Sin. 2015,64(15):154301.

[3] Tao meng. Sound insulation performance of viscoelastic coating which contains horizontal distributed cylindrical holes [C]. International Conference on Materials Engineering and Industrial Applications. Hong Kong,2015:9-14.
[4] Li jingru. Band-gap structure design for reducing vibration and sound and topology optimization of anechoic coating[D].Dalian University of Technology,China.2018.
[5] Panigrahi S N, Jog C S, Munjal M L. Multi-focus design of underwater noise control linings based on finite element analysis[J]. Applied Acoustics. 2008, 69(12): 1141-1153.
[6] Zhao dan, Zhao honggang,etal. A method of enhancing low-frequency and bandwidth of absorption of coatings with cavities[J].The sixteenth conference on underwater ship noise. 2017,549-553.
[7] Zhao yingkun, Sheng meiping, Wang chaojie, et al. Sound insulation properties of underwater honeycomb truss structures[J]. Mechanical Science and Technology,2006,25(6).
[8] Ivansson S M. Anechoic coatings obtained from two and three dimensional monopole resonance diffraction gratings [J].J.Acoust.Soc.Am,2012,131(4).
[9] Zhang chao,Shang dejiang,Zheng fangyuan,et al. Research on sound isolation characteristics of underwater cavity particles layer by FEM[J].Technical Acoustics,2017,36(5).

Micro-scale Aluminium Filled Polymer Composites with Improved Mechanical and Thermal Properties via Hot-pressing

Xiao Zhang[1, a*], Jian Zheng[2,b*], Yahao Liu[2, c], Guibo Yu[2, d], Xiaofeng Zhang[1,e], Zhimin Wang[1, f], Xiang Ma[1, g] and Chunming Zhang[1,h]

[1] Engineering University of PAP, Xi'an, China
[2] Shijiazhuang Campus, Army Engineering University, Shijiazhuang, China

[a]zxleo@foxmail.com; [b]Zhengj2020@163.com; [c] lyh7832@foxmail.com;; [d]panshuo99@126.com; [e]zxf801031@126.com; [f]18700599850@139.com; [g]maxiang1309@126.com; [h] cmz2021@126.com

Abstract: Polymer based thermal interface materials are widely utilized in aeronautical and aerospace equipment, and can significantly contribute to the cooling of devices, which depends on their high thermal conductivities. In this work, a poly(vinylidene fluoride) (PVDF) based composite with micro-scale aluminium particles is fabricated. The aluminium particles were successfully imbedded into the PVDF matrix, and the composite possesses dense structure caused by the hot-pressing process. Benefiting from that, the value of thermal conductivity improves to 1.82 W/(mK), which is 767% higher than that on pristine PVDF. Meanwhile, compared with pristine polymer, the mechanical performances, including tensile strength and modulus, are also significantly improved by 99% and 517%, respectively. The phenomenon can be attributed to the dense inner structure and effective heat transfer pathways of aluminium particles. This simple and effective strategy provides an available method to fabricate the thermoplastic polymeric composite with superior thermal and mechanical properties, used for the high-performance aeronautical and aerospace equipment.
Keywords: Polymer composite, aluminium, Hot-pressing, Mechanical properties, Thermal analysis

1. Introduction

Because of their well performance, such as light mass, flexibility, and easy prepare features, polymeric composites are usually applied in electronic devices, weapon systems, aeronautical and aerospace industry and many other fields[1–3]. Nevertheless, the polymeric materials always have low strength and thermal conductivities, which has limited their applications as the structure requiring superior mechanical and thermal properties, for example thermal interface materials (TIMs)[4, 5]. TIMs are widely applied in the communication, electronic and cooling devices of aeronautical and aerospace equipment, and play a significant role in this field. More exactly, TIMs can significantly contribute to the reliability, lifetime, and high speed of aeronautical and aerospace devices, which largely depend on the thermal dissipation. Polymer-based composites can play a significant role in thermal dissipation [6–8], and are the typical materials as TIMs. It is widely known that the thermal dissipation capability of TIMs mainly depends on their thermal conductivity. Generally, the thermal conductivity of the composite is enhanced by filling fillers with high thermal conductive performance, thereby forming an effective heat transfer pathway in the matrix. To achieve the high thermal conductivity, some kinds of inorganic fillers are embedded into polymer matrices, including metals [9], carbon nanotubes (CNTs) [10], alumina [4], graphene[11], hexagonal boron nitride [12], etc. However, for the nanoscale fillers, because of the strong van der Waals forces, the fillers present agglomeration easily in the polymer matrix.

Aluminium, as a typical metal material, occupies a high thermal conductivity of ~ 237 W/(mK), and has attracted many attentions for serving as thermal conductive material in thermal management field. It has been demonstrated that aluminium provides notable thermal dissipation effectively. Due to its high thermal conductivity, it can be obviously inferred that aluminium can obviously improve the thermal conductivity of polymer composites. Micro-scale aluminium particles, which are easily obtained, has been widely utilized in manufacturing industry [4]. Beside the excellent thermal conductive property, micro-scale aluminium is much more easily to be dispersed in polymer matrix than the fillers with nanoscale size. However, the separated filler particles are hard to construct the heat transfer pathways when they have a large distance. Moreover, because of the large distance between aluminium particles and polymer matrix, caused by the interface incompatibility and weak adhesion, the composite usually has a high interface thermal resistance[4]. Thus, the weight loading of filler should be effectively enhanced, and the fabricated composites should be denser to induce the interface thermal resistance.

Poly(vinylidene fluoride) (PVDF), a kind of fluoropolymer, has been extensively used for the thermoplastic matrix of polymer composite. Owing to its thermoplastic property which, it is very easy to mould this material into a film via the considerably simple methods, for example hot-pressing. Resulting from that, PVDF has been an available choice for fabricating TIMs [10]. Nonetheless, the thermal conductivity of PVDF is extremely low (~0.21 W/(mK)) that it is not suitable for acting as TIM directly. Therefore, an effective strategy should be developed to embed the thermal conductive materials into PVDF appropriately, which can improve the thermal conductivity of composite significantly.

Herein, an aluminium/PVDF composite was carefully fabricated by embedding micro-scale aluminium particles into PVDF matrix via hot-pressing approach. Meanwhile, the thermal properties and mechanical performances were investigated carefully, and both of thermal and mechanical performance of the composites were obviously enhanced, which can be attributed to the dense and effective pathway of heat and stress.

2. Materials and experiments

A. Materials

Aluminium particles with the average micro-scale size of ~48μm were supplied by Aladdin Co., Shanghai, China. Ethyl acetate (≥99.5%) was obtained from Sinopharm Chemical Reagent Co., Ltd., Shanghai, China. Poly(vinylidene fluoride) (PVDF) are provided by Shanghai Dongfu Chemical Technology Co., Ltd.

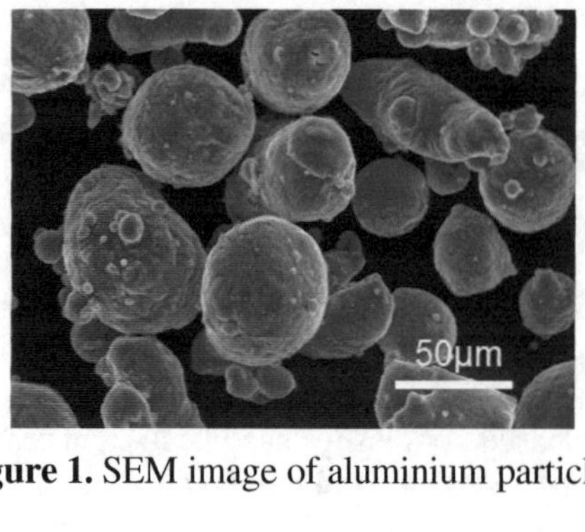

Figure 1. SEM image of aluminium particles.

IOP Conf. Series: Materials Science and Engineering 1102 (2021) 012010 doi:10.1088/1757-899X/1102/1/012010

B. Fabrication of aluminium/PVDF composite

PVDF polymer powder was first dissolved into ethyl acetate with a moderate speed stirring of 200 r/min. The micro-scale aluminium particles were added into the polymer solution with speed stirring for about 10 min. Subsequently, the mixture was under vacuum at 80 °C for 4h to dislodge the ethyl acetate. After that, the obtained composite was pressed under 120 °C with stress of 0.2MPa for 3 hours on hot press machine, and then obtained the final composite.

C. Characterization

The morphology of materials are obtained using scanning electron microscope (SEM, SU-8010, Hitachi). Laser flash diffusivity instrument (DXF-500, TA Instruments) was used to measure thermal diffusivity coefficient. The density was obtained from an automatic density analyzer (PEAB, XS105DU, METTLER TOLEDO, Switzerland). The specific heat capacity is measured via differential scanning calorimetry instrument (DSC Q20, TA Instrument). Tensile tests are conducted on Instron 5980 at tensile speed of 10 mm/min, and the sample possessed the dogbone profile with the tensile length of 12.5mm.

3. Results and analysis

A. Morphologic and Structue of the composite

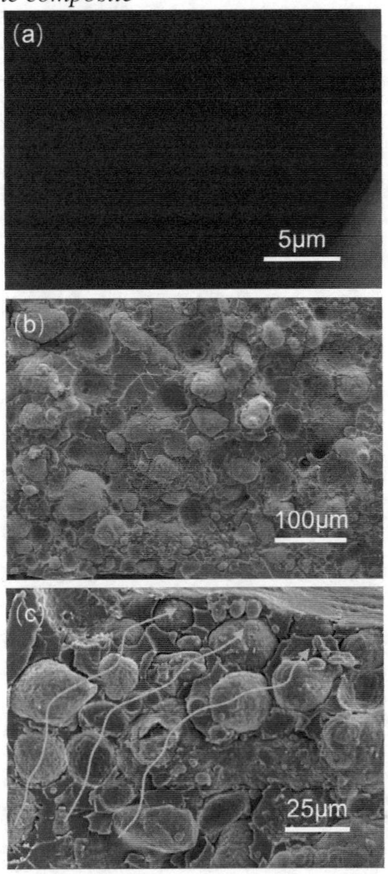

Figure 2. SEM images of (a) pure PVDF; (b) aluminium/PVDF composite; (c) magnification of aluminium/PVDF composite, respectively.

Figure 2 shows the morphologic and structural characterization of the materials used and prepared in this work. Aluminium particles have the micro-scale size, and their size values are not fully uniform, which is beneficial to form denser structural between the big and small particles (Figure 1). As shown in Figure 2a and b, there are clearly different morphologies between pristine PVDF and aluminium/PVDF composite. It is obvious that the aluminium particles are homogeneously dispersed in PVDF matrix without obvious agglomeration. Figure 2c displays the magnification of aluminium/PVDF composite. We can see that the aluminium particles can be driven to be very close to each other, through this hot-pressing process, which can result in effective heat transfer pathways. Due to the higher flowability of PVDF comparing with aluminium particles at high temperature, the aluminium gradually flows towards each other, and they become harder to move than PVDF matrix, which results in the dense aluminium.

B. Thermal conductivity of the composite

TC of pure PVDF and aluminium/PVDF composite were tested, and the values are presented in Figure 3. It can be clearly finded that TC of aluminium/PVDF is much higher than that of pure PVDF. Compared with low thermal conductivity value (0.21 W/(mK)) of PVDF matrix, TC of aluminium/PVDF composite can increase to 1.82 W W/(mK), which is remarkably improved by 767%.

This result can be attributed to the introduction of micro-scale aluminium particles in PVDF and the hot-pressing procedure, which forms the relatively effective path for heat transfer in PVDF matrix. Thus, thermal can conductive along the path from one side to the other side of the composite more effectively, which leads to the improved thermal conductivity.

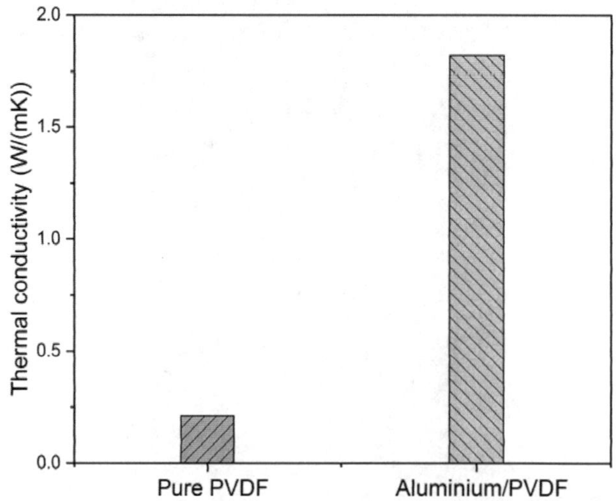

Figure 3. TC of pure PVDF and aluminium/PVDF composite.

C. Mechanical tests

Figure 4 presents the tensile curves of pure PVDF and the composite through tensile measurement at the speed of 10 mm/min. Obviously, the aluminium/PVDF composite exhibits extremely high modulus and tensile strength, which are helpful for the stability of TIMs. More exactly, the Young's and strength are 3.7 MPa and 7.98 MPa, respectively, showing 517% and 99% enhancement, respectively, compared with pure PVDF. More than this, the elongation at break of the aluminium/PVDF is lower than that of pure PVDF, due to the high weight loading of aluminium particles.

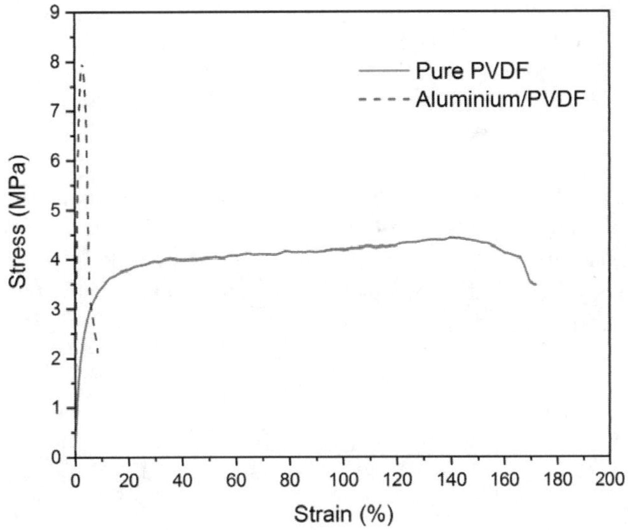

Figure 4. Tensile tests curves of pure PVDF and aluminium/PVDF composite.

4. Conclusion

In summary, the aluminium/PVDF is prepared using micro-scale aluminium particles by hot-pressing approach. TC of the aluminium/PVDF can be enhanced up to 1.82 W/(mK), and is improved by 767%. The elastic modulus and strength are improved by 99% and 517%, respectively. These can be attributed to the dense structure and effective heat transfer pathways of aluminium particles resulting from high weigsht loading and the hot-pressing process. This strategy develops an effective approach to fabricate excellent TIMs for aeronautical and aerospace devices.

References

[1] Zhang X, Zheng J, Fang H, Zhang Y, Bai S 2017 Surface Modified Graphene Oxide Cross-Linking with Hydroxyl-Terminated Polybutadiene Polyurethane: Effects on Structure and Properties, Compos Part A Appl Sci Manuf 103:208-218.

[2] Zhang X, Zheng J, Fang H, Zhang Y, Bai S, He Guan 2018 High Dimensional Stability and Low Viscous Response Solid Propellant Binder Based on Graphene Oxide Nanosheets and Dual Cross-Linked Polyurethane, Compos Sci Technol 161:124-134.

[3] Liu Y, Zheng J, Zhang X, Li K, Du Y, Yu G, Jia Y, Zhang Y 2021 Recent advances on graphene microstructure engineering for propellant-related applications, Journal of Applied Polymer Science 50474.

[4] Zhang X, Zheng J, Fang H, Zhang Y, Bai S, He Guan 2018 Al_2O_3/graphene reinforced bio-inspired interlocking polyurethane composites with superior mechanical and thermal properties for solid propulsion fuel, Compos Sci Technol 167:42-52.

[5] Zhao W, Chen W, Yue Y, Wu S 2017 In-situ two-step Raman thermometry for thermal characterization of monolayer graphene interface material. Appl Therm Eng 113:481–489.

[6] Shen H, Guo J, Wang H, et al 2015 Bioinspired modification of h-BN for high thermal conductive composite films with aligned structure. ACS Appl Mater Interfaces 7:5701–5708.

[7] Zhou T, Wang X, Cheng P, et al 2013 Improving the thermal conductivity of epoxy resin by the addition of a mixture of graphite nanoplatelets and silicon carbide microparticles. Express Polym Lett 7:585–594.

[8] Song WL, Veca LM, Kong CY, et al 2012 Polymeric nanocomposites with graphene sheets-materials and device for superior thermal transport properties. Polymer 53:3910–3916.

[9] Lee W, Kim J 2020 Enhanced through-plane thermal conductivity of paper-like cellulose film with treated hybrid fillers comprising boron nitride and aluminum nitride, Compos Sci Technol 200: 108424.
[10] Zhang X, Zheng J 2019 Superior mechanical and thermal properties of poly(vinylidene fluoride) composites with dense carbon nanotubes, IOP Conference Series: Materials Science and Engineering 479: 012096.
[11] Zhang X, Zheng J 2021 Constructing three-dimensional graphene oxide network by freeze-casting with modified conditions, AIP Conference Proceedings 1995: 020024.
[12] Fang H, Zhang X, Zhao Y, Bai S 2017 Dense graphene foam and hexagonal boron nitride filled PDMS composites with high thermal conductivity and breakdown strength, Compos Sci Technol 152: 43-253.

Dynamic Modelling and Pendulation Laws of Cable Hoisting System

Z X Tian[1,a], X X Lin[1,b,*], J P Li[1,c] and P Hu[1,d]

[1]School of Mechanical Engineering, Shanghai Ocean University, Shanghai, China,

[a]zxtian@shou.edu.cn, [b,*] 15700111533@163.com, [c]li974140159@163.com, [d]15555195015@163.com

Abstract. The dynamic characteristics of the cable hoisting system (CHS) are studied by combining theoretical modelling and visual simulation. The dynamic model is derived using D'Alembert's principle. It contains the effect of horizontal motion and lifting motion, and also takes into account the quality of the cable to improve the accuracy of model. The Runge-Kutta method is used to solve model numerically, and simulation is conducted within Adams environment to verify the dynamic model. Based on the dynamic model and numerical calculation, the effect of horizontal motion, lifting motion and system parameters on swing was studied. The results show that the dynamic model and numerical calculations are accurate; during the movement, the payload oscillation is mainly affected by the initial acceleration, sudden change of acceleration and lifting speed; the length of steel cable and the mass of hoisting payload determine the frequency of CHS, which does not affect the payload oscillation; the oscillation and frequency are affected by lifting laws; and the study of acceleration laws can be used to guide the design and control of CHS.

1. Introduction

High-altitude cable hoisting system (CHS) is to transfer variety of material from one location to another location by the steel cable in order to load and unload huge materials where heavy loads must be moved with extraordinary precision [1]. The CHS acceleration is required for motion and induces undesirable load swing, which poses safety hazards, affects precise positioning, and restricts work efficiency [2, 3]. In order to study the swing of the CHS, it is necessary to establish a dynamic model, plan a more reasonable movement and explore the swing control technology and method. However, due to its underactuated and nonlinear characteristics [4], the system is vulnerable to external interference, indirect control and poor control robustness of the hoisting payload.

Thus, many scholars have done a lot of research on cable crane system, and established two-dimensional two degrees of freedom model of nonlinear system in the early stage. Jaafar, Mohamed, Jamian, Abidin, Kassim and Ab Ghani [5] studied the effect of parameters such as the length of the cable, the mass of the crane payload and the weight of the trolley on the trolley displacement and swing angle of the crane under different voltage inputs. The references [6, 7] only studied the horizontal motion and ignored the lifting movement to control the swing of the lifting load, which has great limitations. Kral, Kreuzer and Wilmers [8] modelled the boom crane as a planar pendulum and the ship as a two-dimensional rigid body undergoing heave, surge, and pitch. The model was used to study the influence of cable hoisting on cargo pendulations. Gao, Chen and Zhang [9] established the nonlinear model of the crane system by using the Lagrange equation method, and cancelled one degree of freedom of the three-dimensional model to obtain two-dimensional crane system dynamic model.

Additionally, some scholars have studied the helicopter suspension system. Lucassen and Sterk [10] were the first to study the helicopter suspension model. The relatively simple model studied the influence of helicopters and lifting systems. This model is limited to three degrees of freedom in the vertical plane, ignoring the aerodynamic force and moment of the lifting weight, so as to analyse the stability of the suspension flight. Finally, the dynamic equation of the suspension was linearized and the data were analysed.

Scholars have studied many researches on overhead crane about the swing characteristics and anti-swing control strategy [11, 12]. However, there are few studies on the movement planner of the CHS [13, 14]. Analysing the effect of the hoisting point movement laws on the hoisting payload swing is conducive to better movement planner, so as to ensure the operating efficiency of lifting and reduce energy consumption.

In this paper, the dynamic model of CHS is derived using D'Alembert's principle, the factors such as cable mass, horizontal motion and lifting motion are considered into the model and the effects of main parameters on pendulation by using the dynamic model are studied.

2. Kinematics of the Cable Hosting System

The dynamic model is of great significance to the exploration of motion laws, the method of swing control and the formulation of operation strategy. Because Adams simulation process is cumbersome and inconvenient to study laws and methods, it is used to verify dynamic models and numerical algorithms. Before establishing the CHS model, the model processing and system motion analysis should be carried out. In order to facilitate the analysis of the dynamic characteristics of the system and the design of effective control methods, the hoisting payload is modeled as a point mass, the steel cable is modeled as an inextensible link.

Basically, the CHS is made up of the hoisting point moving along a horizontal axis with a payload hung from a steel cable, as shown in Figure. 1, where A is the hoisting point, B is the hoisting payload. This work studied the movement of the hoisting payload when the hoisting point moves horizontally and the steel cable moves up or down. Therefore, the coordinate system is established with the hoisting point as the origin of the coordinate system, the direction of hoisting point movement as X axis, and the direction of hoisting payload movement vertically as Z axis. Selecting the position vector of the hoisting point $\boldsymbol{s}$, the position vector of the steel cable $\boldsymbol{l}$, and the position vector of the hoisting payload $\boldsymbol{x}$ as the system degrees of freedom (DOF). It is also convenient for the subsequent simulation verification of Adams. In Figure. 1, the hoisting point is electrically driven by a motor force $\boldsymbol{F}_1$, and the steel cable is electrically driven by a motor force $\boldsymbol{F}_2$.

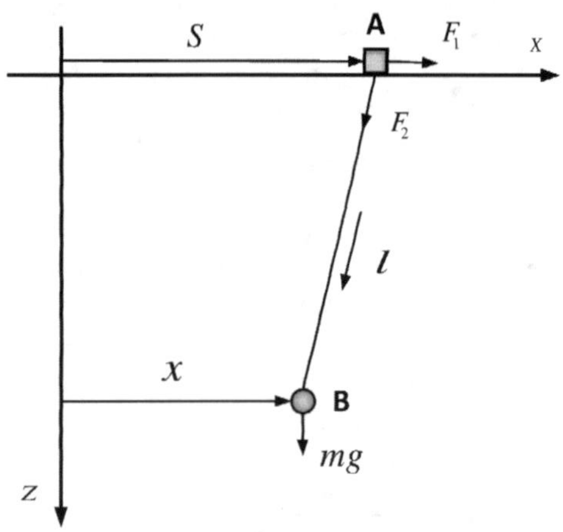

Figure 1. A simple model for two-dimensional cable hoisting system.

The motion of the CHS consists of horizontal motion and rotational motion of the steel cable. As shown in Figure. 1, the position of the hoisting point in coordinate system is given by

$$\begin{cases} x_A = s \\ z_A = 0 \end{cases} \tag{1}$$

The position of the hoisting payload is given by

$$\begin{cases} x_B = x \\ z_B = \sqrt{l^2 - (s-x)^2} \end{cases} \tag{2}$$

The movements of the hoisting cable and rolling cable are different and should be calculated separately. The position of the hoisting cable is given by

$$\begin{cases} x_S = (s+x)/2 \\ z_S = \sqrt{l^2 - (s-x)^2}/2 \end{cases} \tag{3}$$

The position of the rolling cable is considered to be the same as the hoisting point.

The acceleration of the hoisting point, the hoisting payload, and the hoisting cable can be expressed as:

$$\begin{cases} a_{Ax} = \ddot{s} \\ a_{Az} = 0 \\ a_{Bx} = \ddot{x} \\ a_{Bz} = (-(s-x)(l^2-(s-x)^2)\ddot{s} + (s-x)(l^2-(s-x)^2)\ddot{x} + \\ l(l^2-(s-x)^2)\ddot{l} - ((x-s)\dot{l} + l(\dot{s}-\dot{x}))^2)/\sqrt{(l^2-(s-x)^2)^3} \\ a_{Sx} = (\ddot{s}+\ddot{x})/2 \\ a_{Sz} = (-(s-x)(l^2-(s-x)^2)\ddot{s} + (s-x)(l^2-(s-x)^2)\ddot{x} + l(\\ l^2-(s-x)^2)\ddot{l} - ((x-s)\dot{l} + l(\dot{s}-\dot{x}))^2)/(2\sqrt{(l^2-(s-x)^2)^3}) \end{cases} \tag{4}$$

The angle between the cable and the positive direction of Z axis is given by

$$\theta \approx \tan\theta = \frac{s-x}{z_B} \tag{5}$$

The angular acceleration of the steel cable of the center of mass can be expressed as:

$$\begin{aligned} \alpha = (&-l(s-x)(l^2-(s-x)^2)\ddot{l} + l^2(l^2-(s-x)^2)\ddot{s} - l^2(l^2-(s-x)^2)\ddot{x} - 2((\\ &l^2 + (s-x)^2/2)\dot{l} - 3l(\dot{s}-\dot{x})(s-x)/2)(x-s)\dot{l} + l(\dot{s}-\dot{x})/(l^2-(s-x)^2)^{5/2} \end{aligned} \tag{6}$$

According to the above analysis, s, l, and x are the system degrees of freedom, F_1 and F_2 are independent external forces, so the system is an underactuated mechanical system with three degrees of freedom. The relationship between the five factors can be established by the dynamic model. We can observe the movement of the hoisting payload, and analyze the pendulation laws.

3. Dynamic Model of the Cable Hosting System

Since the cable hoisting system is a typical multivariable dynamic system and the upper end of the steel cable as a boundary condition has nonholonomic constraint, the equations of motion of the CHS are derived using D'Alembert's principle, which is a general method to solve the dynamics problems of the system of constrained particle. Its characteristic is that solving the dynamic problems by using the method of studying equilibrium problems of statics. Its equation is given as:

$$\begin{cases}\sum \boldsymbol{F}_i + \sum \boldsymbol{F}_{Ni} + \sum \boldsymbol{F}_{Ii} = \boldsymbol{0} \\ \sum \boldsymbol{M}_O(\boldsymbol{F}_i) + \sum \boldsymbol{M}_O(\boldsymbol{F}_{Ni}) + \sum \boldsymbol{M}_O(\boldsymbol{F}_{Ii}) = \boldsymbol{0}\end{cases} \tag{7}$$

Where, $i = 1,2,\cdots,n$. According to the D'Alembert's principle, we obtain the inertial force and the dynamics of the particle system, so as to establish the dynamic balance equation. The equations of dynamic equilibrium of the CHS can be written as:

$$\begin{cases}\sum F_x = 0, F_1 - F_2(s-x)/l - ma_{Bx} - l\rho a_{sx} - (L-l)\rho a_{Ax} = 0 \\ \sum F_z = 0, F_2 \bullet z_B / l + mg + l\rho g - ma_{Bz} - l\rho a_{sz} = 0 \\ \sum M_c = 0, -F_1(l/2)(z_B/l) + (l/2)(s-x)/lm(g-a_{Bz}) - (l/2)(z_B/l)ma_{Bx} + l^3\rho\alpha/12 = 0\end{cases} \tag{8}$$

Where m is the payload mass, L is the total length of the steel cable, ρ is the linear density of cable.

Equation (8) can be represented by the following matrix-vector form:

$$\begin{bmatrix} M_{11} & M_{12} & M_{13} \\ M_{21} & M_{22} & M_{23} \\ M_{31} & M_{32} & M_{33}\end{bmatrix}\begin{bmatrix}\ddot{x} \\ \ddot{s} \\ \ddot{l}\end{bmatrix} + \begin{bmatrix} H_1 \\ H_2 \\ H_3\end{bmatrix} = \begin{bmatrix} Q_1 \\ Q_2 \\ Q_3\end{bmatrix} \tag{9}$$

Where,

$$\begin{cases} M_{11} = (-\rho l^2 - 2ml)/(2l) \\ M_{12} = (-2\rho Ll + \rho l^2)/(2l) \\ M_{13} = 0\end{cases} \tag{10}$$

$$\begin{cases} M_{21} = -(s-x)(\rho l/2 + m)/(l^2 - (s-x)^2)^{1/2} \\ M_{22} = (s-x)(\rho l/2 + m)/(l^2 - (s-x)^2)^{1/2} \\ M_{23} = -l(\rho l/2 + m)/(l^2 - (s-x)^2)^{1/2}\end{cases} \tag{11}$$

$$\begin{cases} M_{31} = -l^2(\rho l^3/6 + ml^2 - m(s-x)^2)/(2(l^2 - (s-x)^2)^{3/2}) \\ M_{32} = (\rho l^5/6 + m(s-x)^2 l^2 - m(s-x)^4)/(2(l^2 - (s-x)^2)^{3/2}) \\ M_{33} = -l(s-x)(\rho l^3/6 + ml^2 - m(s-x)^2)/(2(l^2 - (s-x)^2)^{3/2})\end{cases} \tag{12}$$

$$\begin{cases} H_1 = 0 \\ H_2 = (l^3(\rho l/2 + m)\dot{x}^2 + l(s-x)^2(\rho l/2 + m)\dot{l}^2 + l^3(\rho l/2 + \\ m)\dot{s}^2 + gl(l^2 - (s-x)^2)(\rho l + m)(l^2 - (s-x)^2)^{1/2} - 2\dot{x}l^3(\rho l/2 \\ +m)\dot{s} - 2l^2(s-x)(\dot{s} - \dot{x})(\rho l/2 + m)\dot{l})/(l(l^2 - (s-x)^2)^{3/2}) \\ H_3 = gm(s-x)/2 + ((s-x)(\rho l^5/3 + \rho(s-x)^2 l^3/6 + m(s-x)^2 l^2 - m(\\ s-x)^4)\dot{l}^2 - 2l(\dot{s} - \dot{x})(\rho l^5/6 + \rho(s-x)^2 l^3/3 + m(s-x)^2 l^2 - m(s-x)^4) \\ \dot{l} + (\rho l^3/2 + ml^2 - m(s-x)^2)l^2(s-x)\dot{s}^2 - 2(l^3\rho/2 + ml^2 - m(s-x)^2)\dot{x}l^2 \\ (s-x)\dot{s} + (\rho l^3/2 + ml^2 - m(s-x)^2)l^2(s-x)\dot{x}^2)/(2(l^2 - (s-x)^2)^{5/2})\end{cases} \tag{13}$$

$$\begin{cases} Q_1 = F_2(s-x)/l - F_1 \\ Q_2 = F_2(l^2-(s-x)^2)^{1/2}/l \\ Q_3 = F_1(l^2-(s-x)^2)^{1/2}/2 \end{cases} \tag{14}$$

It can be seen from the dynamic model that x is related to s , l , F_1 and F_2 . The motion of the hoisting payload can be obtained by planning the movement of the hoisting point and the steel cable, and then the effect of the horizontal movement and the lifting movement on the hoisting payload can be analysed. And the equations of motion of the CHS can be reformed as:

$$\begin{cases} F_1 = 2(M_{31}\ddot{x} + M_{32}\ddot{s} + M_{33}\ddot{l} + H_3)/z_B \\ F_2 = l(M_{21}\ddot{x} + M_{22}\ddot{s} + M_{23}\ddot{l} + H_2)/z_B \\ \ddot{x} = ((((s-x)M_{22} - 2M_{32})/z_B - M_{12})\ddot{s} + ((s-x)M_{23} - 2M_{33})\ddot{l}/z_B \\ +((s-x)H_2 - 2H_3)/z_B - H_1)/(M_{11} - ((s-x)M_{21} - 2M_{31})/z_B) \end{cases} \tag{15}$$

Then the Runge-Kutta method is used to solve a multi degree of freedom vibration equation, and obtain the time domain response curve of vibration. Suppose that:

$$\begin{cases} x_1 = x \\ x_2 = \dot{x} \end{cases} \tag{16}$$

where x_1 is the displacement of the horizontal motion of the hoisting payload, x_2 is the velocity of the horizontal motion of the hoisting payload. After substituting x_1 and x_2 into equation (15), now we can rewrite the equation as:

$$\begin{bmatrix} \dot{x}_1 \\ \dot{x}_2 \end{bmatrix} = \begin{bmatrix} x_2 \\ ((((s-x)M_{22} - 2M_{32})/z_B - M_{12})\ddot{s} + ((s- \\ x)M_{23} - 2M_{33})\ddot{l}/z_B + ((s-x)H_2 - 2H_3)/ \\ z_B - H_1)/(M_{11} - ((s-x)M_{21} - 2M_{31})/z_B) \end{bmatrix} \tag{17}$$

The values corresponding to x_1 and x_2 to each time are obtained by using ode45 module.

4. Implementation, Results and Discussion

The experience of manual operation controls the swing angle within a reasonable range to achieve the hoisting payload positioning and pendulation reduction. Thus, through control of the horizontal movement and the lifting movement, the pendulation can be reduced without affecting the fast and accurate positioning of the CHS. The nonlinear coupling between hoisting traction movement and its swing makes it important to study the effect of the hoisting point movement on hoisting. The pendulation of CHS is related to the acceleration of the hoisting point, the acceleration laws of the hoisting point, the lifting laws of the steel cable, as well as the length of the steel cable and the mass of the hoisting payload, etc. In view of the above aspects, this paper analysed the dynamic model of the CHS and explores its motion laws.

Before analysing the laws of motion, calculating the natural frequency of the CHS, which is similar to a simple pendulum. Since the mass of the steel cable is also taken into account in this system, the frequency is derived on the basis of the system, and the calculation results can be obtained as:

$$L_O = \frac{ml + \frac{1}{2}l \cdot \rho l}{m + \rho l} \tag{18}$$

which is the centre of mass of the CHS. Then the natural frequency of the CHS can be obtained as:

$$f = \frac{1}{2\pi}\sqrt{\frac{(m+\rho l)g}{ml + \frac{1}{2}\rho l^2}} \tag{19}$$

If $m = 2000kg$, $l = 15000mm$, then natural frequency of the CHS f=0.129 . We can judge the vibration of the CHS by analysing the natural frequency and velocity oscillation of the hoisting payload. Since the solution of the dynamic model can obtain two results of the displacement and speed of the hoisting payload, if the speed of the hoisting payload is basically the same as the hoisting point, the system can be considered as stable motion, so the following only analyses the speed of the hoisting payload. In this study, taking a certain garbage grab crane as an example to discuss the pendulation laws of the CHS.

4.1. The Effect of the Acceleration of Hosting Point on Pendulation of the CHS

The movement of the hoisting payload is affected by three factors: horizontal movement, lifting movement and itself swing. In order to express the relationship between the acceleration of the hoisting point and the pendulation more intuitively, the system only performs horizontal motion, and when the speed is a uniform motion, the speed vibration of the hoisting payload is the pendulation of the CHS. The following system parameters were fixed as: $m = 2000kg$, $l = 15000mm$. The acceleration at the starting and stopping moment of the system are respectively $28.125 mm/s^2$, 56.25 mm/s^2 and ∞ , and the speed of the uniform motion in the middle is 225 mm/s .

The speed relationship between the hoisting point and the hoisting payload in Figure. 2 shows that velocity oscillation of payload is affected by the acceleration. The lower acceleration of the hoisting point, the lower velocity oscillation of payload, and when the acceleration is zero, the velocity oscillation of payload remains unchanged. The specific numerical relationship between them in Table 1 shows that when the acceleration is lower to a certain extent, the velocity oscillation of payload is not much affected. For example, when the acceleration is 28.125 mm/s^2 , the velocity oscillation is not obvious. Since the length of the steel cable remains 15000 mm, the frequency is 0.129, which is consistent with the natural frequency of the system.

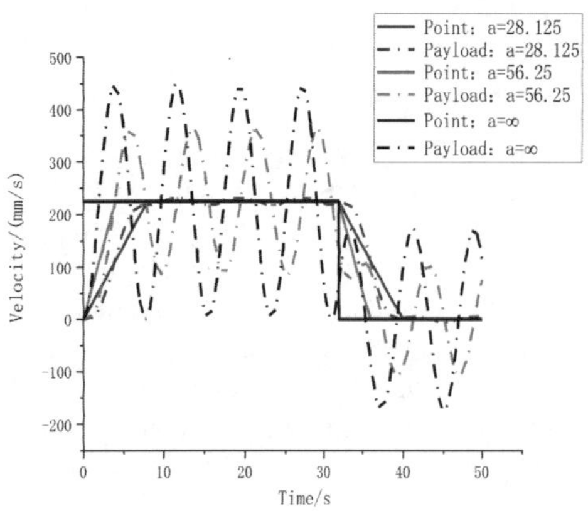

Figure 2. The effect of acceleration of the hoisting point on pendulation.

Table 1. The effect of acceleration on pendulation.

Acceleration (mm/s^2)	velocity Oscillation (mm/s)	Period (s)	Frequency
28.125	6.868	7.76	0.129
37.5	60.673	7.76	0.129
56.25	139.221	7.76	0.129
112.5	201.926	7.76	0.129
∞	225.000	7.76	0.129

4.2. The Effect of the Acceleration Laws of Hosting Point on Pendulation of the CHS

The motion state of the CHS is "stop-start-stop", so the movement mode is to accelerate first and then decelerate. According to the actual situation, three kinds of motion laws that meet the requirements are listed, namely cycloid motion, quadratic motion and simple harmonic motion. They represent the three types of motion laws cycloid, polynomial and trigonometric functions, respectively. Since it has been confirmed that the pendulation of hoisting payload is affected by the acceleration, for the average speed of $225\,mm/s$, the motion distance is different, and the movement laws of the hoisting point has different influences on the pendulation of hoisting payload. Assuming that the system only moves horizontally, the following system parameters were fixed as: $m = 2000kg$, $l = 15000mm$. The motion distances of the hoisting point are $3600\,mm$, $7200\,mm$ and $10800\,mm$, and the corresponding acceleration time is $16\,s$, $32\,s$ and $48\,s$, respectively.

The results are shown in Figure. 3 (a), (b) and (c), respectively. In these figures, when the distance is shorter, the pendulation of hoisting payload affected by the quadratic motion law is the least, but the effect will be amplified in the process of uniform acceleration and deceleration by the increasing distance. Additionally, there is a sudden change of acceleration at the midpoint of the quadratic motion law, and the pendulation phenomenon of the hoisting payload is the most obvious; Compared with the other two motion laws, the cycloidal motion law requires greater acceleration and deceleration in the same time. Thus, with the increasing of the motion distance of the hoisting point, the effect of the pendulation of the hoisting payload is smaller. When the driving distance is $3600\,mm$, the maximum acceleration can reach $45\,mm/s^2$, and the speed changes too fast, leading to obvious pendulation of the hoisting payload. Since the acceleration at the start and stopping moment of the simple harmonic motion law is higher and the acceleration changes suddenly, the hoisting payload is the most obvious pendulation among the three laws, whether it is during the movement or after the stopping movement.

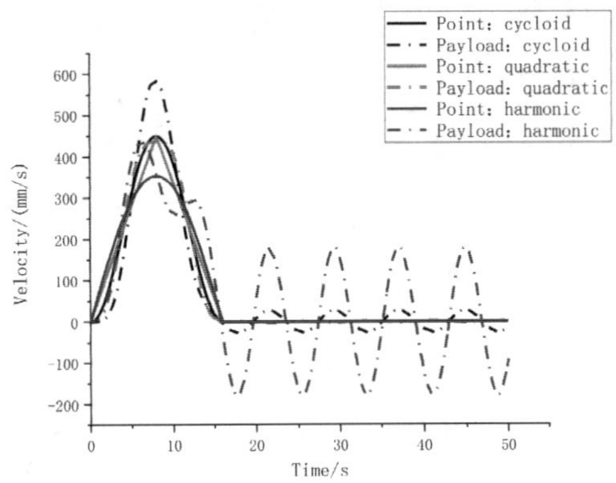

(a)

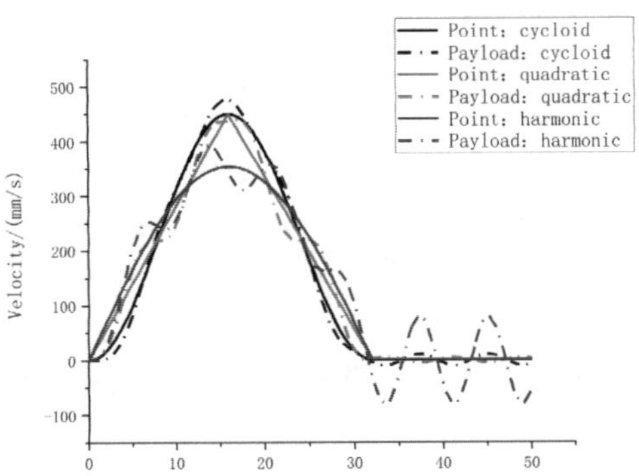

(b)

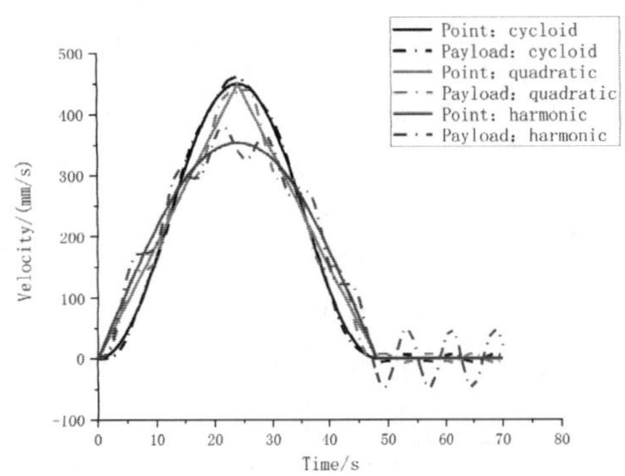

(c)

Figure 3. The effect of acceleration laws of the hoisting point on pendulation.

4.3. The Effect of the Length of Steel Cable and Mass of Hoisting Payload on Pendulation of the CHS

According to (19), the natural frequency is related to the length of the steel cable and the mass of the hoisting payload when the linear density of the cable is constant. Assuming that the hoisting point only moves horizontally, the motion of the hoisting point is uniform motion, which speed is $225\,mm/s$. The mass of the hoisting payload is fixed as: $m = 2000kg$, and the lengths of the steel cables are set as: $l_1 = 1000mm$, $l_2 = 8000mm$ and $l_3 = 15000mm$, respectively. The result of the effect of the length of the hoisting cable on the pendulation of the hoisting payload is shown in Figure. 4. The length of the steel cable is fixed as: $l = 15000mm$, and the mass of the hoisting payload are set as: $m_1 = 20kg$, $m_2 = 200kg$ and $m_3 = 2000kg$, respectively. The result of the effect of the mass of the hoisting payload on the pendulation of the hoisting payload is shown in Figure. 5. The results show that the shorter length of the hoisting cable and the lighter mass of the hoisting payload, the greater frequency. However, the velocity oscillation of payload is not affected by them.

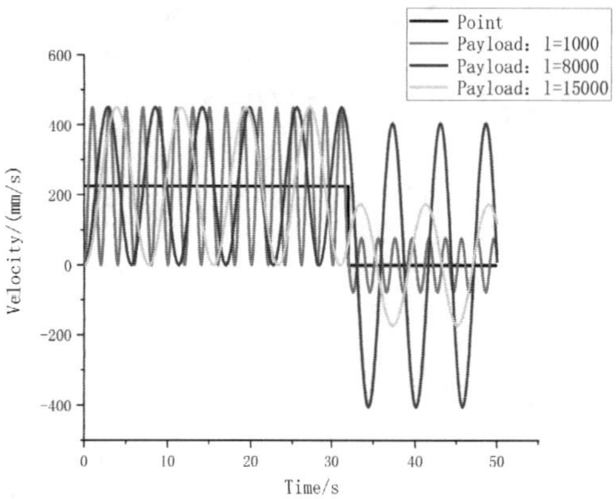

Figure 4. The effect of the length of the hoisting cable on pendulation.

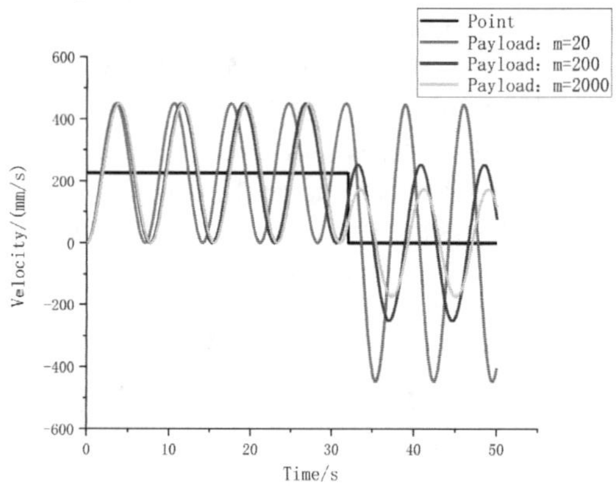

Figure 5. The effect of the mass of the hoisting payload on pendulation.

The specific numerical relationships are shown in Table 2 and Table 3. In these Tables, the frequency is consistent with the natural frequency when the length of the cable is changed, but the frequency is slightly smaller than the natural frequency when the lifting weight is changed. Since the natural frequency is calculated by the centroid of the steel cable, it can be roughly considered that the frequency is consistent with the natural frequency. Since the natural frequency is calculated based on the centre of mass of the CHS, and the mass of the hoisting payload is too small to be close to the mass of the steel cable, resulting in errors. It can be considered that the frequency is consistent with the natural frequency.

Table 2. The effect of the length of the hoisting cable on pendulation.

The Cable Length (mm)	Velocity Oscillation (mm/s)	Period (s)	Frequency
1000	225	2	0.5
8000	225	5.67	0.176
15000	225	7.76	0.129

Table 3. The effect of the mass of the hoisting payload on pendulation.

The Payload Mass (kg)	Velocity Oscillation (mm/s)	Period (s)	Frequency
20	225	7.06	0.142
200	225	7.64	0.131
2000	225	7.76	0.129

4.4. The Effect of the Lifting Laws of the Steel Cable on Pendulation of the CHS

The mass of the hoisting payload is fixed as: $m = 2000kg$, and the motion of the hoisting point is uniform motion, which speed is $225\,mm/s$. The steel cable does lifting motion with the initial length of $15000\,mm$, which are the motion with speed of $0\,mm/s$, upward uniform motion with speed of 225 mm/s and sinusoid motion, respectively. corresponding to three motion states of no lifting motion, uniform rising motion and reciprocating motion. They correspond to three motion states: no lifting motion, uniform lifting motion and reciprocating motion.

The effect of the movement of the CHS on pendulation of the hoisting payload in Figure. 6 shows that the frequency and velocity oscillation of the hoisting payload are different with different lifting motion modes. Since the frequency is affected by the length of the hoisting cable, the frequency of the uniform lifting motion is the fastest, and the frequency of the no lifting movement is the lowest. The velocity oscillation of the uniform lifting motion increases gradually, while that of the sinusoidal motion increases first and then decreases. Since the velocity oscillation is not affected by the length of steel and the acceleration of uniform lifting motion is zero, the velocity of lifting motion will affect the velocity oscillation of the hoisting payload.

In order to further study the relationship between the velocity of lifting motion and the velocity oscillation of hoisting payload, the lifting motion is set to do the uniform lifting motion with the speed of $0\,mm/s$, $112.5\,mm/s$, $225\,mm/s$ and $450\,mm/s$,respectively, and the result in Figure. 7 shows that the larger the velocity of lifting motion, the greater the velocity oscillation of the hoisting payload, and the velocity oscillation increases exponentially with time.

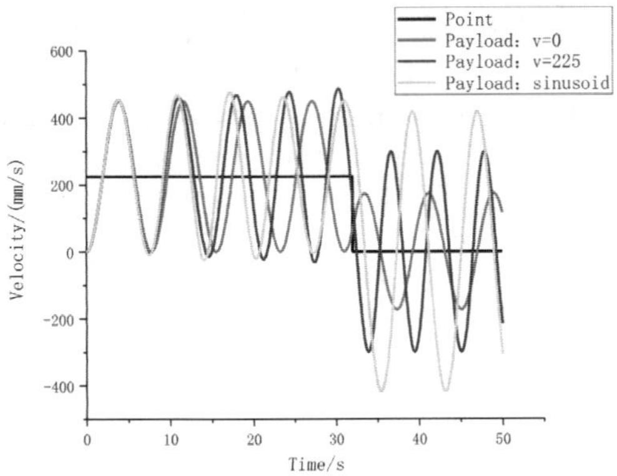

Figure 6. The effect of the lifting laws of the steel cable on pendulation.

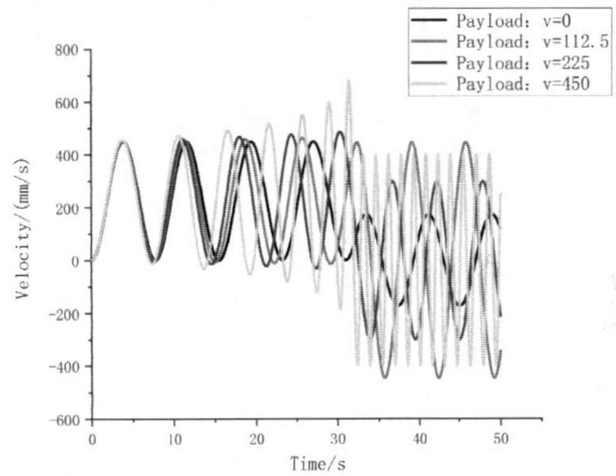

Figure 7. The effect of the velocity of lifting motion of the steel on pendulation.

5. Conclusion

In this paper, the dynamic characteristics of cable crane system are studied by means of theoretical modelling and visual simulation. The dynamic model of two-dimensional cable hoisting system is derived using D'Alembert's principle. In order to lay a foundation for motion planning and reduce pendulation of the CHS, the paper analyses the effect of acceleration laws of the hoisting point, the lifting laws of the steel cable and system parameters on pendulation of the model. Important points of the results have been summarized as follow:

i)The pendulation of the hoisting payload is much affected by the acceleration of the hoisting point. The lower acceleration, the lower pendulation of hoisting payload.

ii)The pendulation of the hoisting payload is affected by the acceleration laws of the hoisting point. Since the pendulation is affected by the acceleration, the quadratic motion can be selected for the short stroke, and the cycloid is the best for the general and long-distance travel.

iii)The frequency of the hoisting payload is affected by the length of the steel cable and the mass of the hoisting payload. The shorter length of the hoisting cable and the lighter mass of the hoisting payload, the greater frequency, which is consistent with the natural frequency.

iv)The pendulation of the hoisting payload is affected by the lifting laws of the steel cable. The frequency is affected by the length of the steel cable and the oscillation is affected by the velocity of lifting motion. The velocity goes up, the pendulation will increase, and the velocity goes down, the pendulation will decrease.

v)After the hoisting point stops moving, the hoisting payload vibrates at the stopping position. The pendulation of the hoisting payload is affected by the velocity, acceleration and frequency at the stopping time.

References

[1] Maghsoudi M, Mohammed Z, Pratiwi A, Ahmad N and Husain A 2012 An experiment for position and sway control of a 3D gantry crane. In: *2012 4th International Conference on Intelligent and Advanced Systems (ICIAS2012)*: IEEE) pp 497-502

[2] Butler H, Honderd G and Van Amerongen J 1991 Model reference adaptive control of a gantry crane scale model *IEEE Control Systems Magazine* **11** 57-62

[3] Ju F, Choo Y and Cui F 2006 Dynamic response of tower crane induced by the pendulum motion of the payload *International Journal of Solids and Structures* **43** 376-89

[4] Liu Y and Yu H 2013 A survey of underactuated mechanical systems *IET Control Theory & Applications* **7** 921-35

[5] Jaafar H I, Mohamed Z, Jamian J, Abidin A F Z, Kassim A M and Ab Ghani Z 2013 Dynamic behaviour of a nonlinear gantry crane system *Procedia Technology* **11** 419-25

[6] Fujioka D, Shah M and Singhose W 2015 Robustness analysis of input-shaped model reference control on a double-pendulum crane. In: *2015 American Control Conference (ACC)*: IEEE) pp 2561-6

[7] Yoon J, Nation S, Singhose W and Vaughan J E 2013 Control of crane payloads that bounce during hoisting *IEEE Transactions on Control Systems Technology* **22** 1233-8

[8] Kral R, Kreuzer E and Wilmers C 1996 Nonlinear oscillations of a crane ship *Zeitschrift für angewandte Mathematik und Mechanik* **76** 5-8

[9] Gao B, Chen H and Zhang X 2006 Dynamic Modeling of a Gantry Crane System *Computer Simulation* **23** 50-2

[10] Lucassen L R and Sterk F J 1965 Dynamic stability analysis of a hovering helicopter with a sling load *Journal of the American Helicopter Society* **10** 6-12

[11] Kim D and Park Y 2017 Tracking control in xy plane of an offshore container crane *Journal of Vibration and control* **23** 469-83

[12] Padula F, Visioli A, Facchinetti D and Saleri A 2015 A dynamic inversion approach for oscillation-free control of overhead cranes. In: *2015 IEEE 20th Conference on Emerging Technologies & Factory Automation (ETFA)*: IEEE) pp 1-6

[13] He B, Fang Y, Liu H and Sun N 2016 Precise positioning online trajectory planner design and application for overhead cranes *Control Theory Appl* **33** 1352-8

[14] Lee H H 2005 Motion planning for three-dimensional overhead cranes with high-speed load hoisting *International Journal of Control* **78** 875-86

Application of Closed Magnetic Circuit Method for Evaluating the Stress Distribution along Depth Direction

Jingyu Di[1,*], Cunfu He[1], Xiucheng Liu[1], and Yung-Chun Lee[2]

[1]College of Mechanical Engineering and Applied Electronics Technology, Beijing University of Technology, Beijing, 100124, China;

[2]Department of Mechanical Engineering, National Cheng Kung University, Taiwan, 70101, China;

[*]B201701018@emails.bjut.edu.cn

Abstract: Electromagnetic nondestructive evaluation method is feasible in assessing stress or residual stress levels in ferromagnetic structures, for there is a strong inherent correlation between the permeability and the stress. This study is based on the closed magnetic circuit method for testing stress distribution along depth. Firstly, a theory is proposed for describing the propagation of electromagnetic energy inside ferromagnetic materials. A layered model is established considering the electromagnetic field experienced different attenuation at different depths. And each layer is assumed to have the identical physical state. Then, multi-frequency eddy current method is proposed to testing the in-depth permeability variation and consequently the stress-depth profile. For simply using the convolution integral, a novel inversion algorithm is given to resolve the convolution integral. To apply the proposed method for testing the stress-depth profile, multi-frequency eddy current experiments are conducted through a four-point bending device; with the specimen material employed is 45# steel. Finally, map the experimental data fitting curve to the theoretical calculation curve, and hence can verify the proposed theory and the inversion algorithm.

Keywords: Multi-frequency eddy current, permeability, stress-depth profile, magnetic circuit.

1. Introduction

The residual stress is generated inside materials after heat treatment or machining being applied to the components [1]-[3]. The residual stress is self-equilibrated inside materials. However, the stress state changes during service period of the components. The tensile stress inside steel structures demonstrates hazards to the load-bearing structures [4]-[6]. Evaluating the distribution of applied stress and residual stress along the depth direction is of importance to avoid sudden failure of the structures. Destructive stress testing methods such as hole-drilling way have high measurement accuracy, but the parts or the components cannot be repaired as original after testing. Nondestructive stress testing methods such as x-ray diffraction and neutron diffraction are expensive and inconvenient for in-situ implementation.

Electromagnetic NDT method is widely used in mechanical properties testing of ferromagnetic materials for inexpensiveness and convenience. The physical principle of electromagnetic NDT is that there are strong inherent correlations between the electromagnetic properties and the microstructures, also the stress state inside materials [7]-[10]. The magnetic property varieties originate from interactions between magnetic domains motion and microstructures such as grain boundary, precipitate, etc., and the stress is also a significant factor influences the magnetizations [11]-[13]. Many researchers focus on developing analytical model to reveal the connections between stresses and magnetic properties. Some attempts have been made to provide a model for stress-depth profile based on spectra and depth-dependent eddy current damping of Barkhausen emissions [14]. Due to high frequency characteristics of Barkhausen noise, most energy generated inside materials is attenuated when reaches the specimen surface, therefore, available testing depth is very limited [15]-[18]. For the stress measurement in components with larger thickness, eddy current method is a promising option. By varying excitation frequencies in the eddy current inducer, the penetration depth of eddy current field can be adjusted. Therefore, multi-frequency eddy current signal carries in-depth permeability information. Due to there is a strong correlation between the change in permeability and the variation in stress, therefore, it is possible to evaluate stress distribution along depth by multi-frequency eddy current method.

In this study, a multi-frequency eddy current closed magnetic circuit model based on convolution integral is employed for illustrating the propagation of electromagnetic field inside materials within stress. To obtain converged solutions of the convolution integral, hypothesis of layered structures along depth is introduced to the model. It is assumed at each layer the permeability (also the stress) is distributed uniformly. Then the proposed inversion algorithm is given for calculating the stress-depth profile. Because the permeability is different at different excitation frequencies, in order to eliminate the difference in permeability caused by frequencies, the variation between the permeability without stress and the permeability under stress is used as the object parameter. Finally, under the existing laboratory condition the four-point bending experiment is implemented, and the test specimen material is of 45# steel. The closed magnetic circuit method based on multi-frequency eddy current is successfully applied to test the stress-depth profile generated by four-point bending experiment. The final experimental results and the theoretical calculated results show a good linear relationship.

2. Methodology

According to the classical electromagnetism theory (1) and (2), which are second-order linear partial differential equations describing the propagation behaviour of the electromagnetic field in free space.

$$\nabla^2 \boldsymbol{E} = \mu_0 \varepsilon_0 \frac{\partial^2 \boldsymbol{E}}{\partial t^2} \tag{1}$$

$$\nabla^2 \boldsymbol{B} = \mu_0 \varepsilon_0 \frac{\partial^2 \boldsymbol{B}}{\partial t^2} \tag{2}$$

Where $\boldsymbol{E}$ and $\boldsymbol{B}$ denote the electric field strength vector and the magnetic induction vector, respectively; the parameters μ_0 and ε_0 represent vacuum permeability and vacuum permittivity, respectively. The electromagnetic field propagation behavior carries information of vacuum permeability and vacuum permittivity. Similarly, if the case comes to the ferromagnetic materials, the electromagnetic field propagating through materials can also reflect the electromagnetic properties of the materials.

Considering the electromagnetic field propagating inside materials is similar to plane wave, and will be attenuated with both distance and frequency. The skin depth expression (3) means when the field goes through distance δ_z, the energy will be reduced to 1/e of its original value.

$$\delta_z = \sqrt{\frac{2\rho}{\omega_e \mu}} \tag{3}$$

Where ω_e is angular frequency of the applied field; ρ is the resistivity of ferromagnetic materials, which is considered to be homogeneous and isotropic; μ is the permeability. For simplicity, the ferromagnetic material without stress is treated to be magnetic isotropy. The magnetization depth is deeper than the skin depth theoretically; therefore, it is feasible to represent the testing depth using the skin depth.

Figure 1 shows the tested material is divided into n layers model; considering each layer of the tested model has the identical stress distribution, thus the ith layer has the same permeability $\bar{\mu}_i$. As sketched in the figure, the different angular frequencies ω_{ei} correspond to the different testing depths z_i, and Δz_i represents the layer thickness. The electromagnetic fields attenuated energy E_i originating from each layer can be estimated as follows (4).

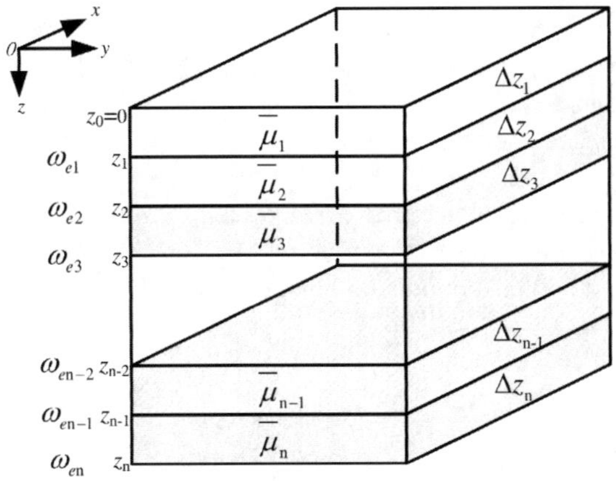

Figure 1. The tested material is divided into n layers equally or unequally, with each layer having identical stress level thus the same permeability.

$$
\begin{aligned}
E_1 &= E_0 e^{-\sqrt{\frac{\omega_e \bar{\mu}_1}{2\rho}}\Delta z_1} \Delta z_1 \\
E_2 &= E_0 e^{-\sqrt{\frac{\omega_e \bar{\mu}_1}{2\rho}}\Delta z_1 - \sqrt{\frac{\omega_e \bar{\mu}_2}{2\rho}}\Delta z_2} \Delta z_2 \\
&\cdots \\
E_i &= E_0 e^{-\sqrt{\frac{\omega_e \bar{\mu}_1}{2\rho}}\Delta z_1 - \sqrt{\frac{\omega_e \bar{\mu}_2}{2\rho}}\Delta z_2 - \cdots - \sqrt{\frac{\omega_e \bar{\mu}_i}{2\rho}}\Delta z_i} \Delta z_i \\
&\cdots \\
E_n &= E_0 e^{-\sqrt{\frac{\omega_e \bar{\mu}_1}{2\rho}}\Delta z_1 - \sqrt{\frac{\omega_e \bar{\mu}_2}{2\rho}}\Delta z_2 - \sqrt{\frac{\omega_e \bar{\mu}_3}{2\rho}}\Delta z_3 - \cdots - \sqrt{\frac{\omega_e \bar{\mu}_n}{2\rho}}\Delta z_n} \Delta z_n
\end{aligned} \tag{4}
$$

Where, E_0 indicates the linear energy density, which is a constant. The detected electromagnetic field total energy E_{energy} on sample surface is the accumulation results of the attenuated electromagnetic fields energy coming from all layers, the expression is as follows (5).

$$
E_{energy} = \sum_{i=1}^{n} e^{-\sum_{j=1}^{i}\sqrt{\frac{\omega_e \bar{\mu}_j}{2\rho}}\Delta z_j} \Delta z_i;\ i = 1,\ 2,\ \ldots,\ n;\ j = 1,\ 2,\ \ldots,\ i \tag{5}
$$

In reality, the permeability $\mu(z)$ of materials is a continuous and smooth function of the depth z. Considering the situation of $\Delta z_i \rightarrow 0$, the integral form can be given as (6).

$$
E_{energy} = \int e^{-\int\sqrt{\frac{\omega_e \mu(z)}{2\rho}}dz} dz \tag{6}
$$

Where, the right of the above formula is a convolution integral term. The permeability of the magnetic materials is very sensitive to stress variation, so the permeability can be expressed as a function of stress; through measuring the permeability variation, the stress level can be tested.

Figure 2 shows the eddy current sensor diagram, employing an excitation coil wound onto a U-shape ferrite core and a testing coil. If electromagnetic field propagates through materials, the integral form of Faraday's law and the integral form of Ampere-Maxwell's law are as follows (7) and (8).

$$\oint_c \boldsymbol{E} \bullet d\boldsymbol{l} = -\frac{d}{dt}\iint_s \boldsymbol{B} \bullet \boldsymbol{n}da \tag{7}$$

$$\oint_c \boldsymbol{H} \bullet d\boldsymbol{l} = I_{free,\ enc} + \frac{d}{dt}\iint_s \boldsymbol{D} \bullet \boldsymbol{n}da \tag{8}$$

Where, the magnetic field strength vector $\boldsymbol{H}$ is equal to $(\boldsymbol{B}/\mu_0-\boldsymbol{M})$; $\boldsymbol{M}$ is the magnetization vector of materials. The displacement vector $\boldsymbol{D}$ is defined as $(\varepsilon_0\boldsymbol{E}+\boldsymbol{P})$; $\boldsymbol{P}$ is the electrode strength vector of materials. $I_{free,\ enc}$ represents the single current. If taking the Ampere integration loop along main flux loop, the derivative of displacement vector flux which is perpendicular to the Ampere integration loop plane is zero.

The magneto-motive force ε_m along a closed magnetic circuit is $\varepsilon_m=\o R_m$, here $\o$ is magnetic flux through cross section. $R_m=L/\mu S$ denotes magneto-resistance calculation formula; L and S represent the length and the cross sectional area of circuit, respectively. The magneto-resistance expression originated from (7) and (8) is given as (9).

$$\varepsilon_m = NI_{free,\ enc} \tag{9}$$

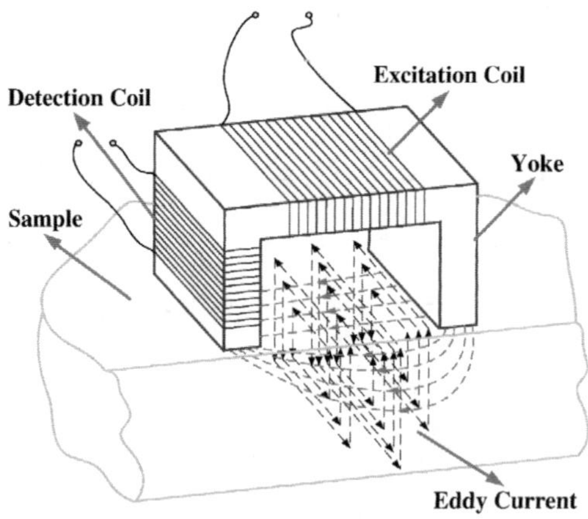

Figure 2. Diagram of the eddy current sensor used, an excitation coil, an induction coil, and a yoke are included.

When an current $I_0(t)=I_0\cos\omega_e t$ is fed into the excitation coil, the voltage induced in induction coil can be obtained as (10).

$$V(t) = \frac{\omega_e nNI_0 \sin\omega_e t}{R_{m,\ total}} \tag{10}$$

Where, $R_{m,\ total}$ denotes the total magnetoresistance of magnetic circuit. According to the continuous convolution formula and electromagnetic propagation attenuation characteristic, letting $I(t)=I_0\sin\omega_e t$, the voltage is obtained as (11).

$$V(t) = \mu_0 nN\omega_e I(t) / (\frac{L_1}{\mu_1 wt} + \frac{L_2}{\mu_2 wt} + \frac{L_3\mu_0\omega_e}{2w\rho}\int e^{-\int\sqrt{\frac{\omega_e\mu(z)}{2\rho}}dz} dz) \tag{11}$$

Where, n is testing coil turns; N is excitation coil turns; L_1, L_2, and L_3 represent the magnetic circuit length in yoke, the air gap length, and the magnetic circuit length in sample respectively; w and t represent the width and thickness of magnetic circuit respectively; μ_1 and μ_2 represent the relative permeability of iron core and air. A discretized inversion algorithm is proposed here to find the solution of $\mu(z)$. For the cases with different testing depths at different frequencies, the convolution integral is simplified as (12).

$$\int e^{-\int\sqrt{\frac{\omega_e \mu(z)}{2\rho}}dz} dz = \sqrt{\frac{2\rho}{\omega_{ei}\bar{\mu}_{1-i}}}(1-e^{-\sqrt{\frac{\omega_{ei}\bar{\mu}_{1-i}}{2\rho}}z_i}) \tag{12}$$

Where, define $\bar{\mu}_{1-i}$ as the averaged permeability of materials from layers 1 to i. Consequently, by solving the corresponding permeability variation at different frequencies based on the proposed discretized inversion algorithm, and using the curve fit tool a continuous solution function $\mu(z)$ can be obtained.

3. Experimental setup and results

The multi-frequency eddy current experimental testing system is shown in Figure 3, including PXI host (built-in signal generation card and data acquisition card), power amplifier, AD/DA modules, the system is controlled by LabVIEW® program. The excitation current can be acquired by connecting 1 Ω sampling resistor in series. Figure 4 shows the geometric size of the U-shaped yoke, with the material employed is silicon steel sheet. The specimen used to generate the stress gradient by the four-point bending device is of 45# steel, with a thickness of 6 mm, here the elastic modulus is about 200 GPa. In this study, five different excitation frequencies are used; ensure the magnetization depth is greater than the testing depth. Table 1 presents the parameters of the closed magnetic circuit.

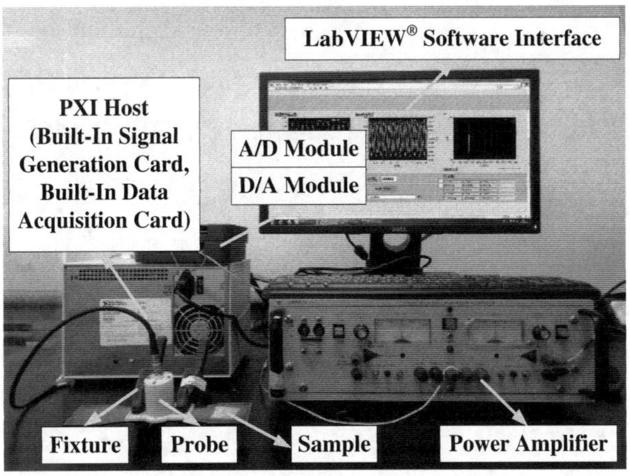

Figure 3. The multi-frequency eddy current experimental testing system setup.

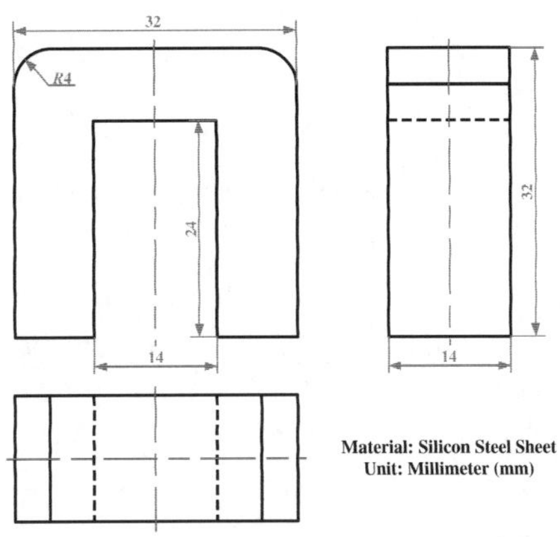

Figure 4. The geometric size of the U-shaped magnetic core, with the material used is silicon steel sheet.

To obtain the measuring signals within the elastic limit, the maximum loading deflection of the sample is 0.25 mm. In a single experiment, synchronously load and record the testing data, and then unload; three measurement signals are collected each. In total, three parallel experimental cycles are implemented, and nine sets of signals are obtained. This study is based on the change of permeability with and without loading; therefore, the experimental results are obtained through the following formula (13).

$$\Delta\bar{\mu}_i \Delta z_i = \int_0^{z_i} \Delta\mu(z)dz - \int_0^{z_{i-1}} \Delta\mu(z)dz \tag{13}$$

Where, Δz_i represents the ith layer thickness, $\Delta\bar{\mu}_i$ denotes the average permeability variation of the ith layer, and $\Delta\mu(z)$ indicates the permeability variation between the loading and the unstressed state.

Table 1. Parameters of the Closed Magnetic Circuit

Symbol	Quantity and Unit	Additional Information
n	300	Testing coil turns
N	150	Excitation coil turns
L_1	80 mm	Length of the magnetic circuit in yoke
L_2	0.18 mm	Length of the air gap
$L_{3,\ 1200Hz}$	15 mm	Magnetic circuit length in sample under 1200 Hz
$L_{3,\ 150Hz}$	16 mm	Magnetic circuit length in sample under 150 Hz
$L_{3,\ 50Hz}$	18 mm	Magnetic circuit length in sample under 50 Hz
$L_{3,\ 10Hz}$	30 mm	Magnetic circuit length in sample under 10 Hz
$L_{3,\ 4Hz}$	48 mm	Magnetic circuit length in sample under 4 Hz
t	8 mm	Thickness of the magnetic circuit
w	14 mm	Width of the magnetic circuit
z	6 mm	Thickness of the sample
ρ	$2.0\times10^{-7}\ \Omega\cdot m$	The resistivity of the sample
μ_0	$4\pi\times10^{-7}$ H/m	The vacuum permeability
μ_1	8000	Relative permeability of the iron core
μ_2	1	Relative permeability of the air

The final experimental results are shown in the following figures. Figure 5 shows the magnetic permeability under different excitation frequencies without and with loading, Figure 6 draws the corresponding permeability variation relative to the unloading situation, and Figure 7 presents the testing depth under different frequencies. More generally, according to the aforementioned algorithm, $N(N+1)/2$ values can be inverted for N different frequencies. Thence 15 permeability variation values can be inverted for 5 frequencies as shown in Figure 8, which the green solid line indicates the theoretical stress gradient, the red circles represent 15 inversion values with error limits, and the black dashed line shows the fitting curve of the inversion values.

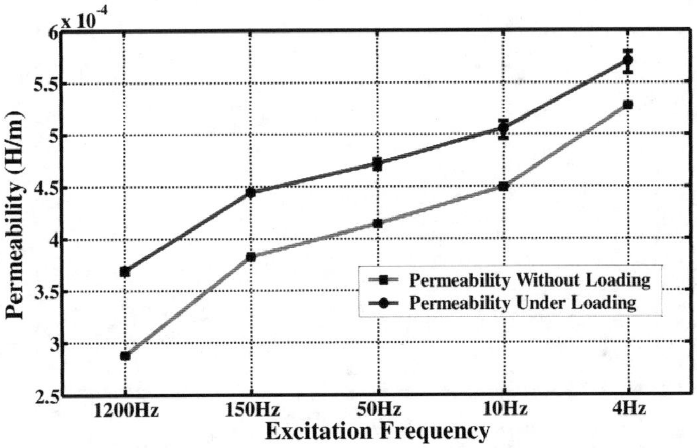

Figure 5. The magnetic permeability under different excitation frequencies without and with loading.

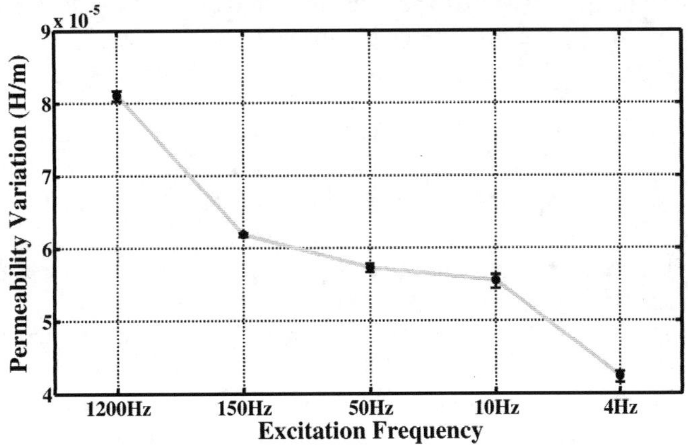

Figure 6. The permeability variation values under different excitation frequencies.

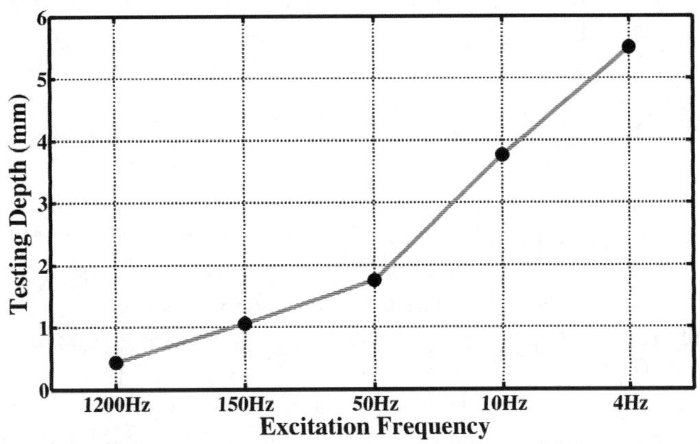

Figure 7. The testing depth values under different excitation frequencies.

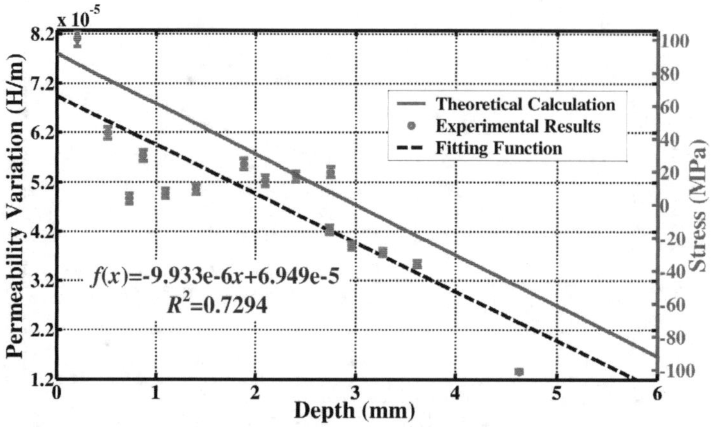

Figure 8. The green solid line indicates the theoretical calculated stress gradient; the red circles represent 15 inversion values based on experimental data with error limits; the black dashed line shows the fitting function of the inversion values.

4. Discussion

In this study, based on the magnetic permeability of ferromagnetic materials changes with stress or residual stress, a new layered model is established and a multi-frequency eddy current sensor to test the stress gradient is designed. The final experimental results present the data fitting curve can map the theoretical calculation curve, and hence can verify the aforementioned theory model and the proposed inversion algorithm. This study provides an effective means for qualitative analysis and approximate quantitative calculation for testing the stress-depth profile.

Using the magnetic circuit method for testing stress distribution along the depth is just an approximate estimation method. Because the iron core permeability is considered to be constant in testing. Due to the magnetic circuit distribution in specimen under different frequencies is extremely complicated; it is almost impossible to use only one-dimensional magnetic circuit length to describe the magnetic circuit distribution in specimen, so the circuit lengths in sample are estimated values. The length of the air gaps is also difficult to measure in each experiment. And the inversion algorithm proposed is only an approximate method. In summary, these all affect the measured permeability values. Therefore, if this assessment method is applied to the actual project, there is still a lot of work to be done.

5. Conclusion

In this research, a layered structure is established for describing the propagation of the electromagnetic energy, and then the multi-frequency eddy current testing principle is described. Assuming each layer has the same physical state, therefore in each layer the permeability is assumed to be identical. According to the hypothetical model, an inversion algorithm is given for simply using the convolution integral. Finally, under the existing condition in laboratory, the four-point bending experiment are carried out. From the experimental results, the data fitting curve can map the theoretical calculation curve. This study provides an effective means for qualitative analysis and approximate quantitative characterization for testing the stress-depth profile. Although using the magnetic circuit method is just an approximate estimation method, this work opens the possibility of testing the stress or residual stress as a function of the depth.

References

[1] O Stupakov. Controllable magnetic hysteresis measurement of electrical steels in a single-yoke open configuration. IEEE Trans. Magn. 2012; 48 (12): 4718-4726.

[2] D J Craik, and M J Wood. Magnetization changes induced by stress in a constant applied field. J. Phys. D. 1970; 3 (7): 1009-1013.

[3] C Jagadish, L Clapham, and D L Atherton. Influence of uniaxial elastic stress on power spectrum and pulse height distribution of surface Barkhausen noise in pipeline steel. IEEE Trans. Magn. 1990; 26 (3): 1160-1163.
[4] O Kypris, I C Nlebedim, and D C Jiles. Mapping stress as a function of depth at the surface of steel structures using a specialized magnetic Barkhausen noise technique. IEEE Trans. Magn. 2012; 48 (11): 4428-4431.
[5] P Vourna, C Hervoches, M Vróna, A Ktena, and E Hristoforou. Correlation of magnetic properties and residual stress distribution monitored by x-ray and neutron diffraction in welded AISI 1008 steel sheets. IEEE Trans. Magn. 2015; 51 (1): 1-4.
[6] D C Jiles, and L Suominen. Effects of surface stress on Barkhausen effect emissions: Model predictions and comparison with x-ray diffraction studies. IEEE Trans. Magn. 1994; 30 (6): 4924-4926.
[7] R W Cahn, and P Haasen. Physical Metallurgy. North Holland Publishing Company, Amsterdam, 1996.
[8] B D Cullity. Introduction to Magnetic Materials. Addison-Wesley, New York, 1972.
[9] D C Jiles, and D L Atherton. Theory of the magnetization process in ferromagnets and its application to the magneto-mechanical effect. J. Phys. D. 1984; 17 (6): 1265-1281.
[10] S Chikazumi. Physics of Ferromagnetism. Oxford University Press, New York, 1997.
[11] A Aharoni. Introduction to the Theory of the Ferromagnetism. Clarendon Press, Oxford, 1996.
[12] D C Jiles. Introduction to Magnetism and Magnetic Materials. Chapman and Hall, New York, 1990.
[13] R C O'Handley. Modern Magnetic Materials: Principles and Applications. John Wiley & Sons, New York, 2000.
[14] O Kypris, I C Nlebedim, and D C Jiles. Barkhausen spectroscopy: Non-destructive characterization of magnetic materials as a function of depth. J. Appl. Phys. 2014; 115 (17): 17E305.
[15] D C Jiles, and S Hariharan. Interpretation of the magnetization mechanism in Terfenol-D using Barkhausen pulse-height analysis and irreversible magnetostriction. J. Appl. Phys. 1990; 67 (9): 5013-5015.
[16] V Moorthy, B A Shaw, and P Hopkins. Surface and subsurface stress evaluation in case-carburised steel using high and low frequency magnetic Barkhausen emission measurements. J. Magn. Magn. Mater. 2006; 299 (2): 362-375.
[17] D J Buttle, C B Scruby, G A D Briggs, and J P Jakubovics. The measurement of stress in steels of varying microstructure by magnetoacoustic and Barkhausen emission. Proc. Roy. Soc. Lond. 1987; 414 (1847): 469-497.
[18] J García-Martín, J Gómez-Gil, and E Vázquez-Sánchez. Non-destructive techniques based on eddy current testing. Sensors. 2011; 11 (3): 2525-2565.

IOP Conf. Series: Materials Science and Engineering 1102 (2021) 012013 doi:10.1088/1757-899X/1102/1/012013

A WEDM-CNC System with Offline Interpolation Post-processing Function

Huliang Ma* **and Yanqing Wang**

School of Mechanical and Vehicle Engineering, Taiyuan University of Technology, Taiyuan, China

*E-mail: mahuliang0006@link.tyut.edu.cn

Abstract. At present, the focus of CNC technology is mainly towards open CNC systems and cloud manufacturing. But the motion control of the CNC system is still mainly composed of interpolation and motion execution. In order to improve the accuracy and quality of interpolation, the applied algorithms are becoming more and more complex. When encountering complex trajectories, minor modifications require complex calculations. To solve this problem, a new CNC system architecture is proposed, that is, adding processing functions after interpolation. Put the minor modification of the trajectory after the interpolation. The article first introduces the CNC architecture with offline interpolation post-processing (OIPP). Then the Interpolation Byte Stream (IBS) format used to represent the interpolation data is introduced. Since the IBS format has the dual characteristics of geometry and string, geometric transformation and text processing both can be used in processing operation. On the basis of the software program and the matching motion controller, the WEDM-CNC system with post-interpolation processing function is implemented. The interpolated data transmission in the developed system adopts a dual-pointer structure, which can effectively control the data transmission. Finally, the feasibility of the new CNC system was verified through machining experiments. Experiments show that offline interpolation post-processing can modify the interpolated data with high precision and flexibility.

1. Introduction

As the CNC technology enters the computer age, there are two ways to process interpolation operations, one is online interpolation and the other is offline interpolation. In the online interpolation mode, interpolation and processing are carried out at the same time, and the obtained data is immediately converted into motor motion. In offline interpolation mode, interpolation and processing are carried out separately, and the toolpath is completely interpolated and then transmitted to the controller.

Online interpolation is also called real-time interpolation, which has been widely studied and applied because it does not need to store a large amount of interpolated data. Offline interpolation needs to store a large amount of data, so it is mainly used for the inspection and simulation of the interpolation method.

Many scholars have explored and improved the structure of the CNC system. Suh and Cheon [1] present a conceptual framework for designing and implementing an intelligent CNC system. Yang and Park [2] had a study on an open WEDM-CNC system with a consideration of the differences between WEDM and NC cutting machines. Han et al [3] developed a PC-based open software-CNC system.

Zhao et al [4] proposed a new model of WEDM-CNC system with digitizer/player architecture. In recent years, the research focus of scholars has shifted to cyber-physical systems [5,6] and cloud manufacturing [7,8]. These literatures mainly research on the overall architecture of the CNC system, and the changes to a single CNC system are not large. They still follow the traditional framework of trajectory planning, interpolation and machining.

With the increase in machining requirements, the complexity of the interpolation algorithm is getting higher and higher. Yeh and Hsu [9] proposed adaptive-feedrate interpolation for parametric curves with a confined chord error. Timar et al [10] presented algorithms for time–optimal control of CNC machines. Zhao et al [11] used a real-time look-ahead interpolation methodology to deal with curvature-continuous B-spline transition scheme. Atmosudiro [12] introduced an interpolation concept for linear blending based on cornu spiral.

These interpolation methods are computationally intensive, and sometimes the CNC developers have to choose between the real-time and the accuracy of the interpolation algorithm. Offline interpolation has the advantage of not needing to consider the real-time of interpolation, so it is more suitable for CNC systems with complex algorithms. In addition, these interpolation methods have higher accuracy for specific objects, and the results of interpolation for other objects are not satisfactory. At the same time, because offline interpolation can get a complete interpolation data result, it also provides the possibility to modify the interpolation result. The modification of the interpolated data is called the offline interpolation post-process (OIPP).

Based on the above, a new architecture of WEDM-CNC system is proposed and shown in figure 1 The dashed line in the figure is the new module.

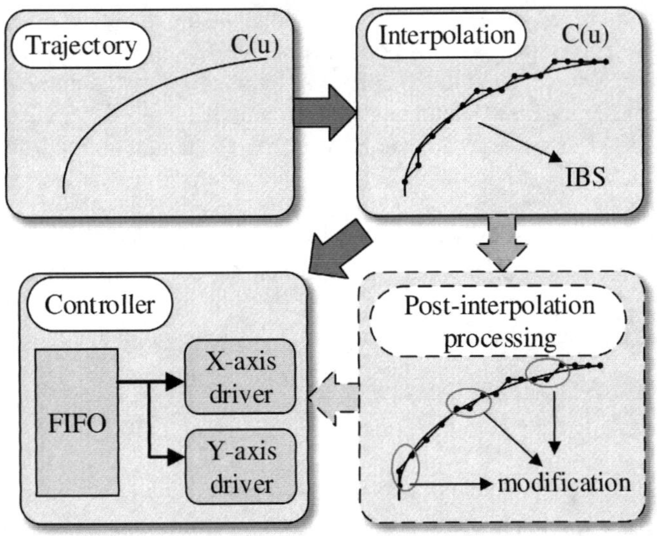

Figure 1. The new model CNC system with post-interpolation processing

The rest of this papar is arranged as follows. Section 2 introduces the content and process of OIPP. Section 3 mainly describes the implementment of the new WEDM-CNC system model. Section 4 demonstrates the feasibility of the new CNC system with a machining example.

2. Offline interpolation post-processing

2.1. Format of Interpolated data

The interpolated data contains every movement of the motor during machining, so it is hundreds of times the trajectory information. As the storage space of the current microcomputer continues to grow, more and more interpolated data can be stored.

In order to adapt to the offline interpolation system, the interpolation data format must meet certain conditions:

- Accurate exercise information.
- Concise expression.
- Readability and modifiability.

Interpolation byte stream (IBS) is a format that meets the above conditions. IBS takes the interpolation byte as the unit, and an interpolation byte unit can contain one or more bytes. The number of bytes is determined by the number and group of axes in the CNC system.

4-axis byte unit	7	6	5	4	3	2	1	0
	DirV	StepV	DirU	StepU	DirY	StepY	DirX	StepX

Figure 2. The 4-axis interpolation byte unit

Table 1. The meaning of bits in byte units

	meaning
Dir X/Y/U/V	Axis movement direction,0 is positive ,1 is negative
Step X/Y/U/V	Axis movement steps,0 is still, 1 is feed

As we all know, the result of interpolation is a single-step movement of each axis. Therefore, one interpolation byte contains the relative movement of each axis. The movement of one axis in the interpolation byte is represented by two bits. One bit indicates the direction of movement, and one bit indicates the number of feed steps. This form of expression refers to the content of literature [4]. After the movement of each axis is arranged in the agreed order, it is converted into byte format, and finally displayed in the form of hexadecimal string on the host computer.

Taking a 4-axis WEDM machine as an example, one byte can contain the information of 4 axes, as shown in figure 2 and table 1. The four axes X/Y/U/V are arranged in one byte in order. In this 4-axis system, one byte unit contains one byte. The hexadecimal string 07h indicates that the UV axes does not move, the Y-axis goes one step forward, and the X-axis reverses one step.

The IBS format can accurately express motion information, and is more concise than the coordinate representation, and has obvious readability and editability. In fact the interpolation format can have multiple representations, as long as the three conditions mentioned above are met.

2.2. Visual editing

The data after interpolation is saved in the computer in the form of IBS string. To modify the data, it must be visualized. The general CNC system usually has a simulation module before machining to check the feasibility of the program and possible interference problems, but it does not provide the function of modifying the local trajectory. The post-interpolation processing provides exactly partial modification function.

For the IBS format, only the starting point coordinates need to be provided, and the movement of each motor can be obtained according to the movement information of each byte unit, and converted into the coordinates of the tool or the platform.

For machine tool operators, visual verification is required when generating and modifying trajectories. This is because although the motor movement corresponding to a single byte is very simple in the post-interpolation processing, the number of interpolation bytes is very large, and visualization tools must be used.

2.3. Post- processing

The OIPP function is the operation after the interpolation. It is the same as trajectory planning in that both are operations on trajectories. The difference is that the trajectory planning faces the machining size requirements of the workpiece, and the OIPP faces the interpolated data.

The OIPP faces many linear segments. Therefore, almost geometric operations on linear segments can be applied to post- processing. At the same time, because it is a string format, some file operations can also be performed.

The geometric transformation includes functions such as rotation, mirroring and zoom, as shown in figure 3. Rotation and mirroring operations are common operations in machining. The rotation operation needs to regenerate the IBS according to the rotated coordinates, but if the rotation angle is a multiple of 90°, it can be directly replaced according to the corresponding byte unit. The most commonly used processing is the mirroring operation of the relative axis, which can be directly processed according to the text. The zoom operation can be used when the accuracy allows.

The text editing operation has functions such as cutting, pasting and arraying, as shown in figure 4. The cutting operation divides the IBS into two sections at the specified position. The obtained two line segments can be used to combine with other line segments. The pasting operation is to combine two IBS segments into one. Array operation is to copy one IBS into multiple and combine into one. Text editing operations can handle IBS more flexibly.

In machining, when the operator encounters a certain position of the trajectory that needs slight modification, or adds certain features, the text editing operation can be used. In fact, OIPP is to provide operators with a way to directly access the final motion. After mastering the principle of OIPP, the operator can fully manipulate the movement of all motors.

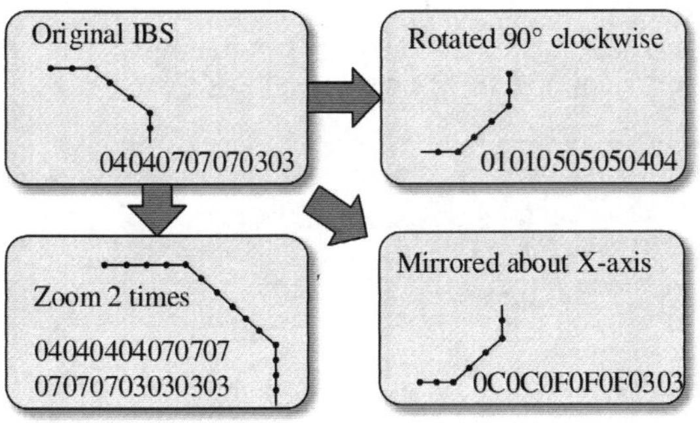

Figure 3. Geometric transformation operation to IBS

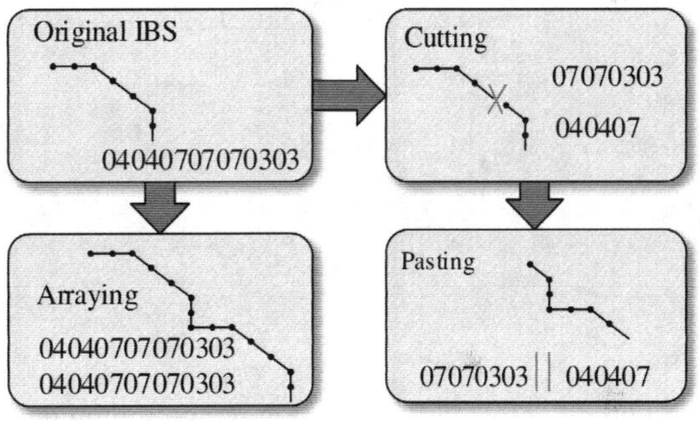

Figure 4. Text editing operation to IBS

3. Implementation with the new model

The interpolated data modified by OIPP is different from the original toolpath information and cannot be converted to toolpath information for storage, so it must be sent to the motion controller in the suitable format to execute the movement. The IBS format is in bytes, so it is very easy to transmit and decode. In this section, an IBS-based WEDM-CNC system is implemented to prove the feasibility of the new architecture.

3.1. Basic structure of the controller

The amount of data after interpolation is very large, and the controller resources are limited and it is impossible to download completely at one time. In order to meet the real-time processing, a buffer module must be set in the controller. FIFO memory has unique first-in first-out characteristics, suitable for storing and reading processing data. The signal interface of the FIFO memory is shown in figure 5. FIFO memory is used as a data buffer module, receiving interpolation data and sending it to the subsequent execution module according to the signals. The *almost_full* and *almost_empty* signals of the FIFO are sent to the parameter control module to transmit the status of the FIFO.

The data is stored in the FIFO and read as needed. The read data is converted into motor motion control signal. The other functional modules are similar to the existing controller functions and will not be detailed one by one. The main modules of the controller is shown in figure 6. In addition to the interpolation data sending module, there are also position feedback and parameter control modules.

The position feedback module sends the position signal of the motors or gratings to the computer after processing. The parameter control module includes a parameter controller and a speed controller. The parameter controller receives the parameters and sends them to the corresponding interface. The speed controller adjusts the feed speed according to the speed control information of the parameter controller and the machining state of the machine tool.

WEDM-CNC belongs to low-speed machining. In order to maintain the gap of spark discharge, it is necessary to receive gap status information and adjust the speed at any time. The current CNC system usually contains a speed control module. This article implements an adjustable speed control module for FIFO memory.

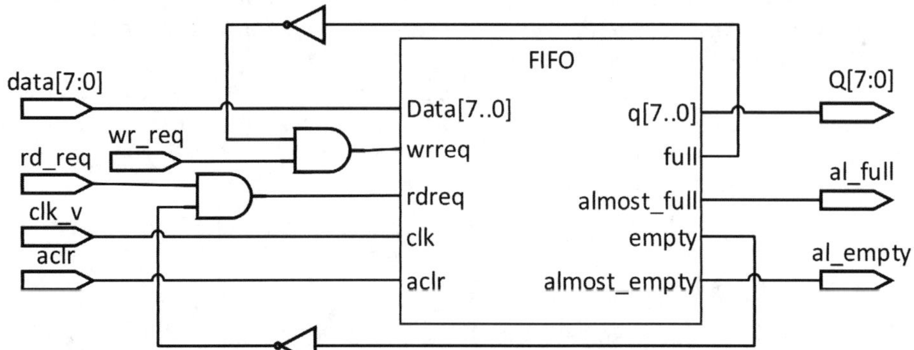

Figure 5. The FIFO memory interface circuit

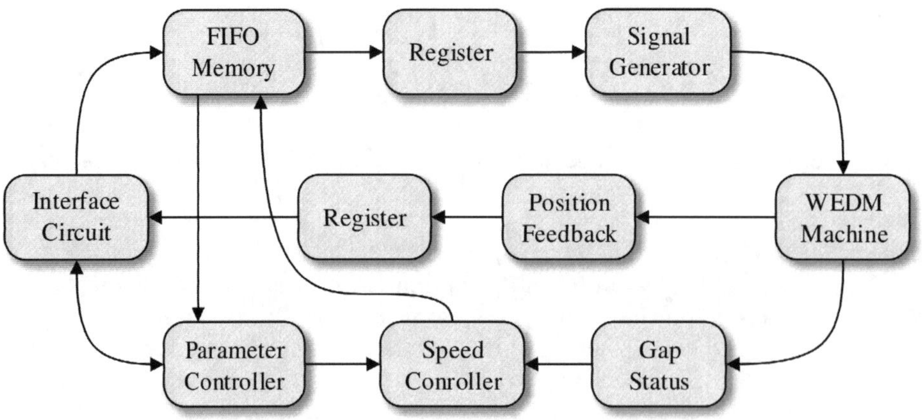

Figure 6. The main modules of the controller

3.2. The interpolated data flow structure

The large amount of interpolated data is stored in Industrial Personal Computer (IPC), and the IPC software adopts a dual pointer structure, including sending pointer (SP) and receiving pointer (RP). The sending pointer points to the position after sending the buffer block, and the receiving pointer points to the feedback after the motor moves. The dual pointer method can not only send forward, but also send in reverse when the discharge gap is short-circuited.

The interpolated data flow of the new WEDM-CNC system is shown in figure 7. The IBS module sends data to the controller in the form of a fixed-length buffer block each time it sends data. The data length when receiving is subject to the actual number of feed steps. At the same time, the status signal of the FIFO is checked regularly, and the next set of buffer blocks is sent when the data in the FIFO is almost empty. The dual pointer structure can send forward and reverse data streams, implementing forward or reverse machining.

The other parts of the CNC system in the figure are consistent with the existing system.

4. Applications

In order to verify the feasibility of the new WEDM-CNC system model, the computer software and new controller of the new CNC system were developed and installed on the existing WEDM machine tools for processing experiments. The experimental equipment is shown in figure 8.

The controller adopts FPGA chip, and the hardware program design is compiled and loaded into the chip. The machine tool is a high-speed wire cutting machine, and the movement is driven by stepper motors. The machining accuracy is 1μm.

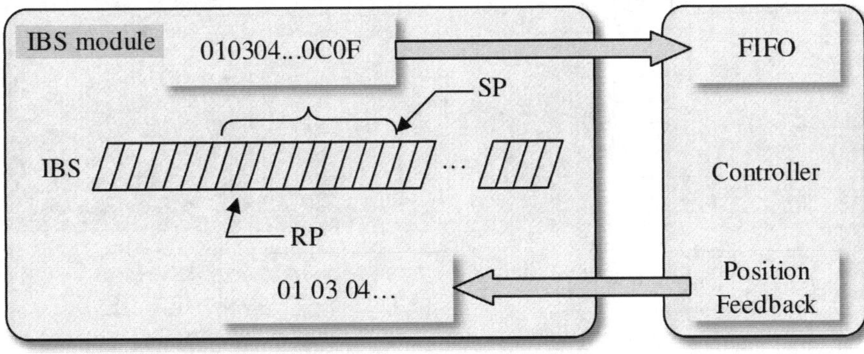

Figure 7. The interpolated data flow structure

The original and modified IBS are visualized as shown on the left of figure 9. The workpiece machined by the new WEDM-CNC system is shown on the right of figure 9. As shown in the figure, the five corners of the five-pointed star are processed into different shapes through post-interpolation processing. Cut, paste, rotate and mirror operations are used. After replacing the sharp corners of the five-pointed star with a different IBS, the processed trajectory is obtained.

Figure 10 shows the trajectory with two redundant parts after interpolation, resulting in corresponding defects during machining. After using the cutting method in OIPP, the redundant parts are removed and the machining results are also improved.

Figure 8. The experimental equipment

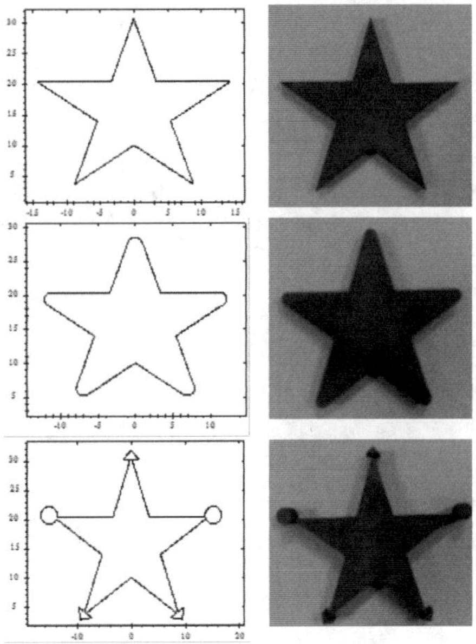

Figure 9. IBS visualization and machined workpieces

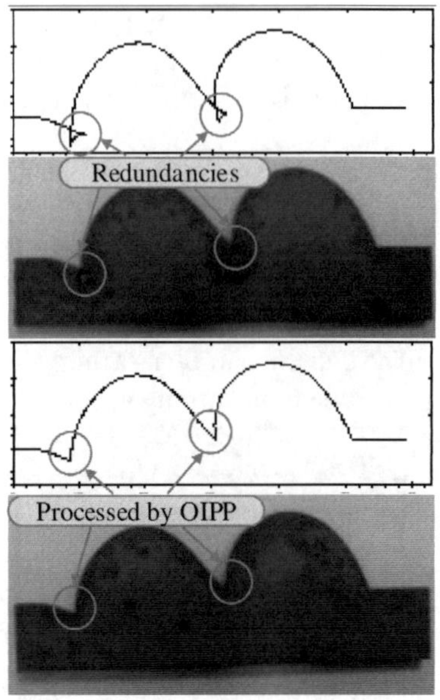

Figure 10. Curve processed by OIPP and its machined workpieces

The CNC system developed in section 3 is used for machining, and the machined workpieces are consistent with the graphics displayed by the software. Data transmission is timely during machining, and the machining efficiency is equivalent to that of the original machine.

5. Conclusions

A WEDM-CNC system with OIPP function was proposed and implemented in WEDM machine tools. The feasibility of OIPP has been verified. The interpolation data format can also be optimized to achieve better storage and transmission characteristics. OIPP is to perform surgical processing on the interpolated data, which can be modified according to the conditions of the machine tool to achieve micron-level or sub-micron-level accuracy. OIPP has great application prospects for high-precision nonlinear machining.

The OIPP is applied to the CNC system of offline interpolation, so that the offline interpolation is not limited to the inspection and simulation of the interpolation method, but the CNC system architecture with huge advantages.

Acknowledgments

This work is funded by the Shanxi Key R & D Program Project under grant number 201903D121048.

References

[1] Suh S H and Cheon S U 2002 A framework for an intelligent CNC and data model *The International Journal of Advanced Manufacturing Technology* **19** pp 727-735

[2] Yang M Y and Park J H 2002 A study on an open architecture CNCsystem with a NURBS interpolator for WEDM *The International Journal of Advanced Manufacturing Technology* **19** pp 664-668

[3] Han Z Y, Wang Y Z and Fu H Y 2007 Development of a PC-based open architecture software-CNC system *Chinese Journal of Aeronautics* **20** pp 272-281

[4] Zaho W S, Zheng J M, Xi X C, et al 2016 A new model of WEDM-CNC system with digitizer/player architecture *Procedia CIRP* **42** pp 210-214

[5] Wang L H, Torngren M, and Onori M 2015 Current status and advancement of cyber-physical systems in manufacturing *Journal of Manufacturing Systems* **37** pp 517-527

[6] Deng C Y, Guo R F, Zheng P, et al 2018 From Open CNC Systems to Cyber-Physical Machine Tools: A Case Study *Procedia CIRP* **72** pp 1270-1276
[7] Zhong R Y, Wang L H and Xu X 2017 An IoT-enabled Real-time Machine Status Monitoring Approach for Cloud Manufacturing *Procedia CIRP* **63** pp 709-714
[8] Xu X 2012 From cloud computing to cloud manufacturing *Robotics and computer-integrated manufacturing* **28** pp 75-86
[9] Yeh S S and Hsu P L 2002 Adaptive-feedrate interpolation for parametric curves with a confined chord error *Computer-aided design* **34** pp 229-237
[10] Timar S D, Farouki R T, Smith T S and Boyadjieff C L 2005 Algorithms for time–optimal control of CNC machines along curved tool paths *Robotics and Computer-Integrated Manufacturing* **21** pp 37–53
[11] Zhao H, Zhu L, Ding H 2013 A real-time look-ahead interpolation methodology with curvature-continuous B-spline transition scheme for CNC machining of short line segments *The International Journal of Machine Tools and Manufacture* **65** pp 88–98
[12] Atmosudiro A, Verl A, Lechler A and Morán X 2016 An Interpolation Concept for Linear Blending Based on Cornu Spiral *Procedia CIRP* **46** pp 348–351

Effects of Misalignment on the Sliding Rates of Parallel Line Gear Pair

Yifan Lin[1] and Yangzhi Chen[1, a]

[1]School of Mechanical and Automotive Engineering, South China University of Technology, Guangzhou 510641, Guangdong, China

[a]E-mail: meyzchen@scut.edu.cn

Abstract. The sliding rates of parallel line gear pair (PLGP) are equal to zero and pure rolling meshing can be achieved under ideal condition. In this paper, effects of misalignment on the sliding rates of PLGP were studied. Firstly, the geometric model of PLGP was presented. Then the contact curves of PLGP with misalignment were obtained by the tooth contact analysis (TCA) method and the sliding rates were calculated. The calculation results of the example show that PLGP under ideal condition or with only axial deviation accomplishes transmission along the theoretical contact curves, the sliding rates are equal to zero, and pure rolling meshing can be achieved. For PLGP with centre distance deviation and parallelism misalignment, the sliding rates are not equal to zero and the form of relative motion between the meshing surfaces combines rolling and sliding. This investigation provides a theoretical basis for the further study of meshing efficiency and wear prediction of line gear pair under actual operating condition.

1. Introduction

Relative sliding between the meshing surfaces occurs throughout the meshing process (except for the pitch point) of the commonly used gear pairs [1, 2]. Sliding rates are often used as a measure of the relative sliding between the meshing surfaces [3]. Researchers study the relative sliding of a gear pair by calculating its sliding rates. Xu et al. [4] derived the formula of sliding rates of a novel cycloidal gear pair and analysed the effects of various parameters on the sliding rates. Wang et al. [5] presented an internal gear pair with a high contact ratio and studied its sliding rates. Relative sliding between the meshing surfaces under dry friction and starved-oil conditions will lead to problems such as low meshing efficiency and severe wear, which can limit the application of the gear pairs.

Line gear (LG), a novel gear mechanism invented on the basis of space curve meshing theory, accomplishes transmission via point contact meshing of a pair of space conjugate curves named the driving and driven contact curves [6, 7]. By selecting the design parameters of a line gear pair, the sliding rates can be equal to zero, which means that relative sliding during the meshing process can be eliminated and pure rolling meshing can be achieved theoretically. For example, when selecting cylindrical helix as the driving contact curve of a parallel line gear pair (PLGP), the sliding rates are equal to zero throughout the meshing process [8].

However, misalignment under actual operating condition can change the meshing positions of PLGP, which might result in relative sliding between the meshing surfaces and change the sliding rates. In this paper, the effects of misalignment on the sliding rates of PLGP were studied. Firstly, the geometric model of PLGP was presented. Then the contact curves of PLGP with misalignment were obtained by the tooth contact analysis (TCA) method. Finally, the sliding rates were calculated and the effects of misalignment on the sliding rates were analysed through a calculation example.

2. Sliding rates of PLGP

2.1. Geometric model of PLGP

The geometric model of PLGP is shown in Figure 1. $o-xyz$ and $o_f-x_fy_fz_f$ are fixed coordinate systems. Coordinate systems $o_1-x_1y_1z_1$ and $o_2-x_2y_2z_2$ are fixed with the driving LG and driven LG respectively. The driving LG and driven LG rotate around axis z_1 and axis z_2 respectively with a centre distance of a. When the theoretical driving contact curve $\boldsymbol{r}_1$ is a right-hand cylindrical helix, the theoretical driven contact curve $\boldsymbol{r}_2$ is a left-hand one according to the space curve meshing theory [6, 7]. k is the instantaneous meshing point. The expressions of $\boldsymbol{r}_1$ and $\boldsymbol{r}_2$ in $o_1-x_1y_1z_1$ and $o_2-x_2y_2z_2$ are shown as Equation (1) and Equation (2) respectively:

$$\boldsymbol{r}_1^{(1)}(t_1)=\begin{pmatrix}\mathrm{m}_1\cos(t_1)\\ \mathrm{m}_1\sin(t_1)\\ \mathrm{n}_1(t_1+\pi)\end{pmatrix}\quad\left(t_{1s}\le t_1\le t_{1e}\right) \tag{1}$$

$$\boldsymbol{r}_2^{(2)}(t_2)=\begin{pmatrix}\mathrm{m}_2\cos(t_2)\\ \mathrm{m}_2\sin(t_2)\\ \mathrm{n}_2t_2\end{pmatrix}\quad\left(\frac{t_{1s}}{i_{12}}\le t_2\le\frac{t_{1e}}{i_{12}}\right) \tag{2}$$

where, subscript $i=1$ and $i=2$ represent the parameters of the driving LG and driven LG respectively. m_i is the helix radius of $\boldsymbol{r}_i$, $\mathrm{m}_1>0$ and $\mathrm{m}_2>0$. n_i is the pitch parameter of $\boldsymbol{r}_i$, $\mathrm{n}_1>0$ and $\mathrm{n}_2<0$. i_{12} is the transmission ratio, $\mathrm{m}_2=i_{12}\mathrm{m}_1$ and $\mathrm{n}_2=-i_{12}\mathrm{n}_1$. t_i is the parameter indicating the scope of $\boldsymbol{r}_i$. $\boldsymbol{r}_1$ and $\boldsymbol{r}_2$ start to mesh at t_{is} and separate at t_{ie}.

In the normal plane of an arbitrary point on $\boldsymbol{r}_i$, an arc $\mathbf{P}_i$ is selected as the generatrix and made to sweep along $\boldsymbol{r}_i$, and the tooth surface $\boldsymbol{\Sigma}_i$ is formed. The expressions of $\boldsymbol{\Sigma}_1$ and $\boldsymbol{\Sigma}_2$ in $o_1-x_1y_1z_1$ and $o_2-x_2y_2z_2$ are shown as Equation (3) and Equation (4) respectively:

$$\boldsymbol{\Sigma}_1^{(1)}(t_1,\theta_1)=\begin{pmatrix}\cos(t_1)(\mathrm{m}_1-R_1(\cos(\theta_1+\phi_{z1})+\sin(\phi_{z1})))+\mathrm{n}_1R_1\sin(t_1)(\sin(\theta_1+\phi_{z1})-\cos(\phi_{z1}))(\mathrm{m}_1^2+\mathrm{n}_1^2)^{-1/2}\\ \sin(t_1)(\mathrm{m}_1-R_1(\cos(\theta_1+\phi_{z1})+\sin(\phi_{z1})))-\mathrm{n}_1R_1\cos(t_1)(\sin(\theta_1+\phi_{z1})-\cos(\phi_{z1}))(\mathrm{m}_1^2+\mathrm{n}_1^2)^{-1/2}\\ \mathrm{n}_1(t_1+\pi)+\mathrm{m}_1R_1(\sin(\theta_1+\phi_{z1})-\cos(\phi_{z1}))(\mathrm{m}_1^2+\mathrm{n}_1^2)^{-1/2}\end{pmatrix} \tag{3}$$

$$\boldsymbol{\Sigma}_2^{(2)}(t_2,\theta_2)=\begin{pmatrix}\cos(t_2)(\mathrm{m}_2-R_2(\cos(\theta_2+\phi_{z2})+\sin(\phi_{z2})))+\mathrm{n}_2R_2\sin(t_2)(\sin(\theta_2+\phi_{z2})-\cos(\phi_{z2}))(\mathrm{m}_2^2+\mathrm{n}_2^2)^{-1/2}\\ \sin(t_2)(\mathrm{m}_2-R_2(\cos(\theta_2+\phi_{z2})+\sin(\phi_{z2})))-\mathrm{n}_2R_2\cos(t_2)(\sin(\theta_2+\phi_{z2})-\cos(\phi_{z2}))(\mathrm{m}_2^2+\mathrm{n}_2^2)^{-1/2}\\ \mathrm{n}_2t_2+\mathrm{m}_2R_2(\sin(\theta_2+\phi_{z2})-\cos(\phi_{z2}))(\mathrm{m}_2^2+\mathrm{n}_2^2)^{-1/2}\end{pmatrix} \tag{4}$$

where, θ_i is the parameter indicating the scope of $\mathbf{P}_i$. R_i is the radius of $\mathbf{P}_i$. ϕ_{zi} is the modification angle of the tooth profile.

$\mathbf{n}_{10}$ and $\mathbf{n}_{20}$ are the unit normal vectors of $\boldsymbol{\Sigma}_1$ and $\boldsymbol{\Sigma}_2$, of which the expressions in $o_1-x_1y_1z_1$ and $o_2-x_2y_2z_2$ are shown as Equation (5) and Equation (6) respectively:

$$\mathbf{n}_{10}^{(1)}(t_1,\theta_1)=\left(\frac{\partial\boldsymbol{\Sigma}_1^{(1)}}{\partial\theta_1}\times\frac{\partial\boldsymbol{\Sigma}_1^{(1)}}{\partial t_1}\right)\Bigg/\left|\frac{\partial\boldsymbol{\Sigma}_1^{(1)}}{\partial\theta_1}\times\frac{\partial\boldsymbol{\Sigma}_1^{(1)}}{\partial t_1}\right| \tag{5}$$

$$\mathbf{n}_{20}^{(2)}(t_2,\theta_2)=\left(\frac{\partial\boldsymbol{\Sigma}_2^{(2)}}{\partial t_2}\times\frac{\partial\boldsymbol{\Sigma}_2^{(2)}}{\partial\theta_2}\right)\Bigg/\left|\frac{\partial\boldsymbol{\Sigma}_2^{(2)}}{\partial t_2}\times\frac{\partial\boldsymbol{\Sigma}_2^{(2)}}{\partial\theta_2}\right| \tag{6}$$

The direction from the tooth entity to the outside of the tooth is the positive direction of $\mathbf{n}_{10}$, while the direction from the outside of the tooth to the tooth entity is the positive direction of $\mathbf{n}_{20}$.

2.2. Contact curves with misalignment

Misalignment under the actual operating condition can change the meshing positions, thus the actual contact curves might deviate from the theoretical ones. As shown in Figure 2, misalignment of PLGP is defined as: centre distance deviation, namely, an extra displacement δa in the direction of axis x_f; axial deviation, namely, an extra displacement δb in the direction of z_f; parallelism misalignment of the two rotating shafts, namely, the angle γ_h between the projections of axis z_1 and axis z_2 in the plane $y_f o_f z_f$ (crossed axes misalignment), and the angle γ_v in the plane $x_f o_f z_f$ (intersected axes misalignment). The values of δa, δb, γ_h and γ_v shown in Figure 2 are all positive.

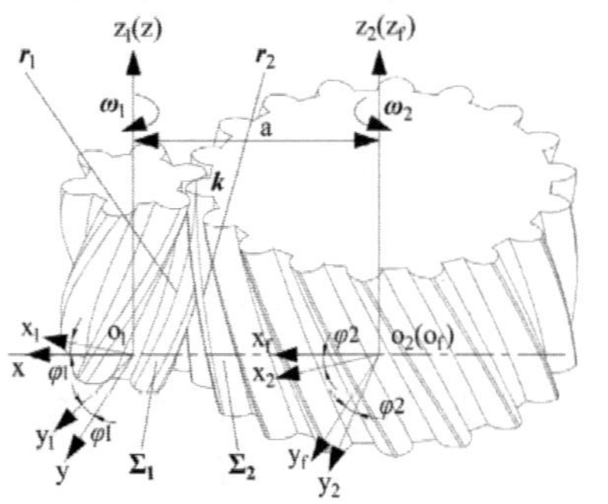

Figure 1. Geometric model of PLGP.

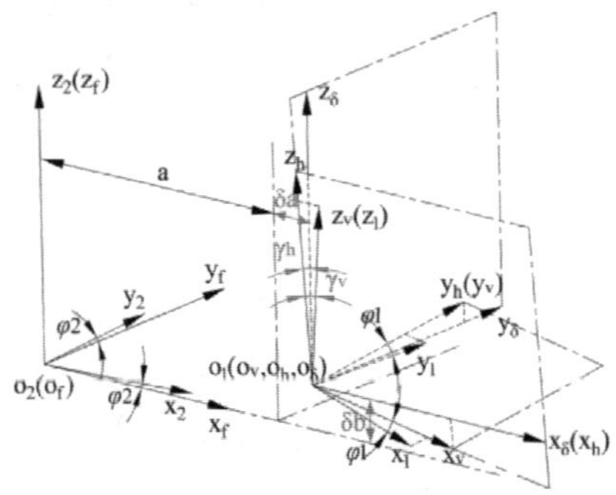

Figure 2. Misalignment of PLGP.

Transformation matrix M_{f1} from $o_1 - x_1 y_1 z_1$ to $o_f - x_f y_f z_f$ and M_{f2} from $o_2 - x_2 y_2 z_2$ to $o_f - x_f y_f z_f$ are shown as Equation (7) and Equation (8) respectively:

$$\mathrm{M}_{\mathrm{f}1} = \begin{pmatrix} \cos(\gamma_v)\cos(\varphi_1) & \cos(\gamma_v)\sin(\varphi_1) & \sin(\gamma_v) & a+\delta a \\ \sin(\gamma_h)\sin(\gamma_v)\cos(\varphi_1)-\cos(\gamma_h)\sin(\varphi_1) & \sin(\gamma_h)\sin(\gamma_v)\sin(\varphi_1)+\cos(\gamma_h)\cos(\varphi_1) & -\sin(\gamma_h)\cos(\gamma_v) & 0 \\ -\cos(\gamma_h)\sin(\gamma_v)\cos(\varphi_1)-\sin(\gamma_h)\sin(\varphi_1) & -\cos(\gamma_h)\sin(\gamma_v)\sin(\varphi_1)+\sin(\gamma_h)\cos(\varphi_1) & \cos(\gamma_h)\cos(\gamma_v) & \delta b \\ 0 & 0 & 0 & 1 \end{pmatrix} \quad (7)$$

$$\mathrm{M}_{\mathrm{f}2} = \begin{pmatrix} \cos(\varphi_2) & -\sin(\varphi_2) & 0 & 0 \\ \sin(\varphi_2) & \cos(\varphi_2) & 0 & 0 \\ 0 & 0 & 1 & 0 \\ 0 & 0 & 0 & 1 \end{pmatrix} \quad (8)$$

where φ_1 and φ_2 are the rotational angles of the driving LG and driven LG respectively.

Utilizing coordinate transformation, Σ_i and $\mathbf{n}_{i0}$ in $o_f - x_f y_f z_f$ are obtained as Equation (9) and Equation (10):

$$\begin{cases} \Sigma_1^{(\mathrm{f})}(t_1,\theta_1,\varphi_1) = \mathrm{M}_{\mathrm{f}1}(\varphi_1)\cdot\Sigma_1^{(1)}(t_1,\theta_1) \\ \Sigma_2^{(\mathrm{f})}(t_2,\theta_2,\varphi_2) = \mathrm{M}_{\mathrm{f}2}(\varphi_2)\cdot\Sigma_2^{(2)}(t_2,\theta_2) \end{cases} \quad (9)$$

$$\begin{cases} \mathbf{n}_{10}^{(\mathrm{f})}(t_1,\theta_1,\varphi_1) = \mathrm{L}_{\mathrm{f}1}(\varphi_1)\cdot\mathbf{n}_{10}^{(1)}(t_1,\theta_1) \\ \mathbf{n}_{20}^{(\mathrm{f})}(t_2,\theta_2,\varphi_2) = \mathrm{L}_{\mathrm{f}2}(\varphi_2)\cdot\mathbf{n}_{20}^{(2)}(t_2,\theta_2) \end{cases} \quad (10)$$

where L_{f1} and L_{f2} are the 3-order submatrices of M_{f1} and M_{f2} respectively.

The TCA equations at the meshing points is expressed as Equation (11) [9]:

$$\begin{cases} \mathbf{\Sigma}_1^{(f)}(t_1,\theta_1,\varphi_1) = \mathbf{\Sigma}_2^{(f)}(t_2,\theta_2,\varphi_2) \\ \mathbf{n}_{10}^{(f)}(t_1,\theta_1,\varphi_1) = \mathbf{n}_{20}^{(f)}(t_2,\theta_2,\varphi_2) \end{cases} \tag{11}$$

A meshing period from meshing starting to meshing ending of a pair of teeth of the PLGP is divided evenly into K meshing positions. Rotational angle of the driving LG is given as φ_{1k} ($k=1,...,K$), and the other parameters t_{1k}, t_{2k}, θ_{1k}, θ_{2k} and φ_{2k} can be obtained by solving Equation (11). (t_i,θ_i) represents the coordinate of the meshing point on $\mathbf{\Sigma}_i$. $\theta_1=\theta_1(t_1)$, $\theta_2=\theta_2(t_2)$ and $t_2=t_2(t_1)$ can be obtained by fitting the TCA results. Submitting $\theta_1=\theta_1(t_1)$ into Equation (3), the driving contact curve with misalignment can be obtained as $\boldsymbol{r}_{1ac}=\boldsymbol{r}_{1ac}(t_1)$. Submitting $\theta_2=\theta_2(t_2)$ and $t_2=t_2(t_1)$ into Equation (4), the driven contact curve with misalignment can be obtained as $\boldsymbol{r}_{2ac}=\boldsymbol{r}_{2ac}(t_1)$.

2.3. Sliding rates of PLGP

PLGP accomplishes transmission via point contact meshing of a pair of contact curves $\boldsymbol{r}_{1ac}$ and $\boldsymbol{r}_{2ac}$. As shown in Figure 3, $\boldsymbol{r}_{1ac}$ and $\boldsymbol{r}_{2ac}$ mesh at point k initially. After a period of Δt, point k_1 on $\boldsymbol{r}_{1ac}$ meshes with point k_2 on $\boldsymbol{r}_{2ac}$. Δs_1 and Δs_2 are the arcs on $\boldsymbol{r}_{1ac}$ and $\boldsymbol{r}_{2ac}$ respectively when Δt approaches zero. Sliding rates of $\boldsymbol{r}_{iac}$ are defined as the limit value of the ratio of the length difference between Δs_1 and Δs_2 to the length Δs_i, as expressed in Equation (12) and Equation (13):

$$\sigma_1 = \lim_{\Delta s_1 \to 0} \frac{\Delta s_1 - \Delta s_2}{\Delta s_1} = \frac{\mathrm{d}s_1/\mathrm{d}t_1 - \mathrm{d}s_2/\mathrm{d}t_1}{\mathrm{d}s_1/\mathrm{d}t_1} = 1 - \frac{\mathrm{d}s_2/\mathrm{d}t_1}{\mathrm{d}s_1/\mathrm{d}t_1} \tag{12}$$

$$\sigma_2 = \lim_{\Delta s_2 \to 0} \frac{\Delta s_2 - \Delta s_1}{\Delta s_2} = \frac{\mathrm{d}s_2/\mathrm{d}t_1 - \mathrm{d}s_1/\mathrm{d}t_1}{\mathrm{d}s_2/\mathrm{d}t_1} = 1 - \frac{\mathrm{d}s_1/\mathrm{d}t_1}{\mathrm{d}s_2/\mathrm{d}t_1} \tag{13}$$

$\boldsymbol{r}_{1ac}$ and $\boldsymbol{r}_{2ac}$ can be denoted as $\boldsymbol{r}_{1ac}=[x_1(t_1)\ \ y_1(t_1)\ \ z_1(t_1)]^T$ and $\boldsymbol{r}_{2ac}=[x_2(t_1)\ \ y_2(t_1)\ \ z_2(t_1)]^T$, then $\mathrm{d}s_1/\mathrm{d}t_1=((x_1'(t_1))^2+(y_1'(t_1))^2+(z_1'(t_1))^2)^{0.5}$, $\mathrm{d}s_2/\mathrm{d}t_1=((x_2'(t_1))^2+(y_2'(t_1))^2+(z_2'(t_1))^2)^{0.5}$. Thus Equation (12) and Equation (13) can be rewritten as:

$$\sigma_1 = 1 - \frac{((x_2'(t_1))^2+(y_2'(t_1))^2+(z_2'(t_1))^2)^{0.5}}{((x_1'(t_1))^2+(y_1'(t_1))^2+(z_1'(t_1))^2)^{0.5}} \tag{14}$$

$$\sigma_2 = 1 - \frac{((x_1'(t_1))^2+(y_1'(t_1))^2+(z_1'(t_1))^2)^{0.5}}{((x_2'(t_1))^2+(y_2'(t_1))^2+(z_2'(t_1))^2)^{0.5}} \tag{15}$$

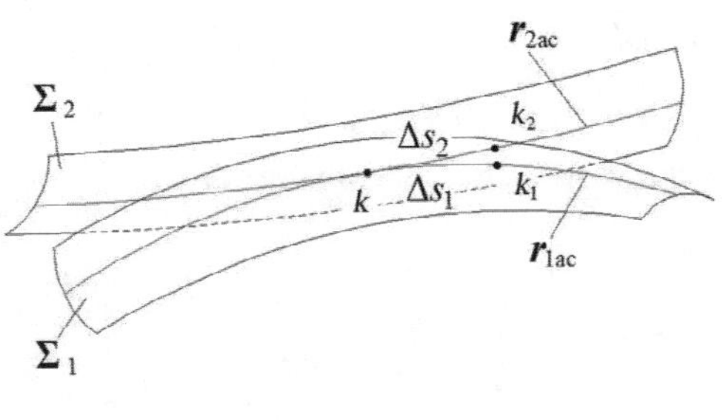

Figure 3. Relative sliding between the $\boldsymbol{r}_{1ac}$ and $\boldsymbol{r}_{2ac}$.

3. Calculation example and analysis

A PLGP was given as an example. The design parameters and misalignment values of the PLGP are given in Table 1 and Table 2 respectively.

Table 1. Design parameters of PLGP.

Parameters	Driving LG	Driven LG
(m_i, n_i) /mm	(10, 20)	(30, -60)
R_i /mm	4	4
ϕ_{zi} /°	30	150
(t_{si}, t_{ei})	(-π, -π/2)	(0, -π/6)
Transmission ratio	3	

Table 2. Misalignment values of PLGP.

misalignment	values
Centre distance deviation /mm	$\pm 0.01,\ \pm 0.02$
Axial deviation /mm	$\pm 0.01,\ \pm 0.02$
Crossed axes misalignment /°	$\pm 0.01,\ \pm 0.02$
Intersected axes misalignment /°	$\pm 0.01,\ \pm 0.02$

The sliding rates of PLGP were calculated by the method mentioned in Section 2. The sliding rates under ideal condition can be obtained as $\sigma_1 = \sigma_2 = 0$, which means that PLGP under ideal condition accomplishes transmission along the theoretical contact curves and pure rolling meshing can be achieved throughout the meshing process.

The sliding rates with different kinds of misalignment are shown from Figure 4 to Figure 7. As shown in Figure 4, for PLGP with only axial deviation, the sliding rates are also equal to zero throughout the meshing process, which means that PLGP still accomplishes transmission along the theoretical contact curves and the form of relative motion between the meshing surfaces is pure rolling.

However, as shown from Figure 5 to Figure 7, when the two rotating shafts of PLGP have centre distance deviation or parallelism misalignment, the contact curves deviate from the theoretical ones, the sliding rates are not equal to zero, and the relative motion between the meshing surfaces combines both rolling and sliding. As shown in Figure 5, $\sigma_1 > 0$ and $\sigma_2 < 0$ when $\delta a > 0$, while $\sigma_1 < 0$ and $\sigma_2 > 0$ when $\delta a < 0$. σ_1 and σ_2 keep constant during the meshing process. As shown in Figure 6, $\sigma_1 > 0$ and $\sigma_2 < 0$ when $\gamma_h > 0$, while $\sigma_1 < 0$ and $\sigma_2 > 0$ when $\gamma_h < 0$. σ_1 and σ_2 approximately keep constant during the meshing process. As shown in Figure 7, $\sigma_1 > 0$ and $\sigma_2 < 0$ when $\gamma_v > 0$, while $\sigma_1 < 0$ and $\sigma_2 > 0$ when $\gamma_v < 0$. $|\sigma_1|$ and $|\sigma_2|$ increase approximately linearly from meshing starting to meshing ending. For this calculation example, $|\sigma_i|$ varies from approximately 0.00008 to 0.00014 when $|\gamma_v| = 0.01^\circ$ and from approximately 0.00015 to 0.00028 when $|\gamma_v| = 0.02^\circ$. The absolute values of the sliding rates increase with the increase of the absolute values of these three kinds of misalignment.

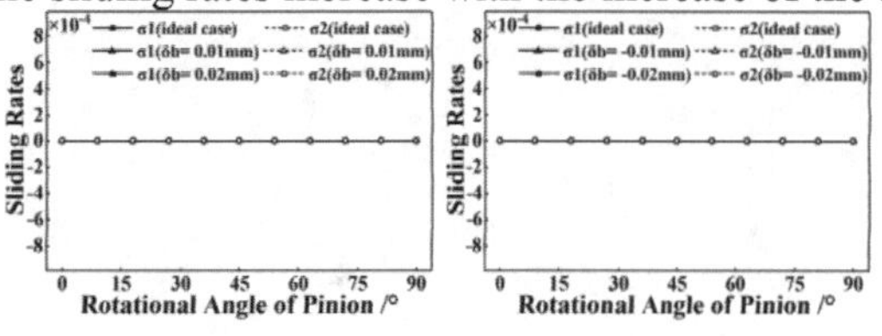

(a) $\delta b > 0$. (b) $\delta b < 0$.

Figure 4. Sliding rates with only axial deviation

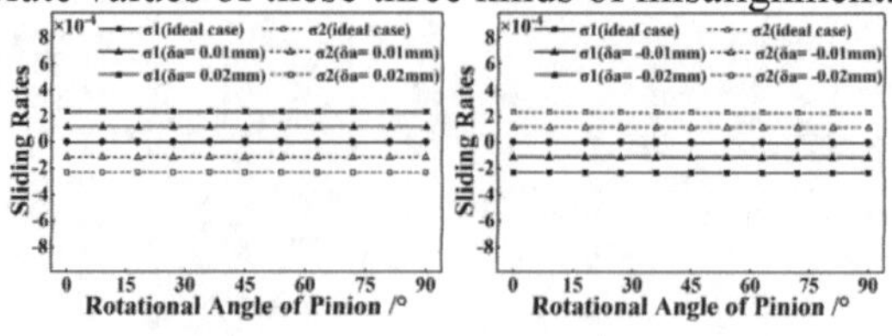

(a) $\delta a > 0$. (b) $\delta a < 0$.

Figure 5. Sliding rates with only centre distance deviation

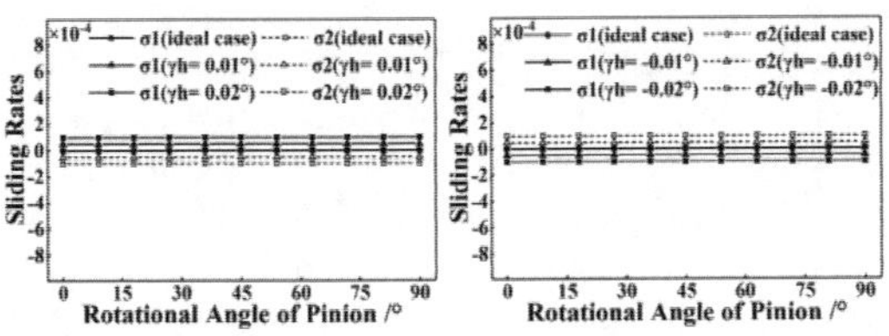

(a) $\gamma_h > 0$. **(b)** $\gamma_h < 0$.

Figure 6. Sliding rates with only crossed axes misalignment

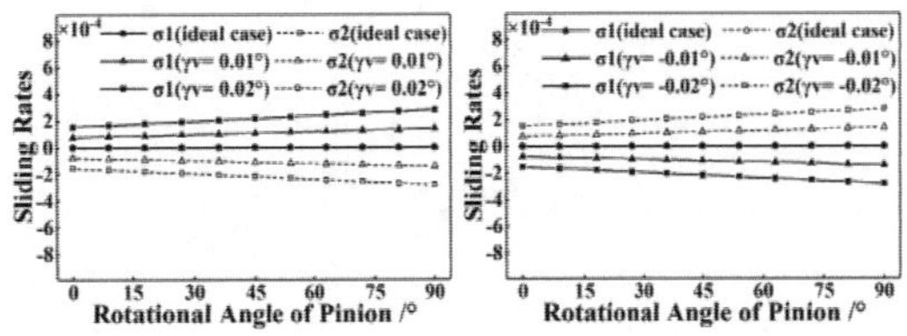

(a) $\gamma_v > 0$. **(b)** $\gamma_v < 0$.

Figure 7. Sliding rates with only intersected axes misalignment

4. Conclusions

By analysing the effects of misalignment on the sliding rates of PLGP, the main results can be concluded as follows:

1. PLGP under ideal condition or with only axial misalignment accomplishes transmission along the theoretical contact curves, the sliding rates are equal to zero throughout the meshing process, and pure rolling meshing can be achieved.

2. PLGP with centre distance deviation and parallelism misalignment accomplishes transmission along a pair of contact curves deviating from the theoretical ones, the sliding rates are not equal to zero, and the relative motion between the meshing surfaces combines both rolling and sliding. The sliding rates of PLGP with intersected axes misalignment vary during the meshing process, which indicates that intersected axes misalignment can affect the stability of the gear transmission.

3. The sliding rates of PLGP are close to zero, which indicates approximate pure rolling meshing can be achieved under actual operating condition.

Acknowledgements

The authors gratefully acknowledge the financial support from the National Natural Science Foundation of China (No. 51575191) and 2019 Guangzhou Technology Project (No. 201904010368).

References

[1] Xu H, Kahraman A, Anderson N E, Maddock D G. Prediction of mechanical efficiency of parallel-axis gear pairs. Journal of Mechanical Design 129(1): 58-68 (2007).

[2] Xu H, Kahraman A. A frictional efficiency loss model for helical gears. 17th International Conference on Design Theory and Methodology. Power Transmission and Gearing Conference, Long Bench, California, USA, 2005: 897-908.

[3] Litvin F L. Theory of gearing. Washington,D.C (USA): NASA Reference Publication, 1989.

[4] Xu L Z, Li S, Wang W P. Sliding ratio for novel cycloidal gear drive. Proceedings of the Institution of Mechanical Engineers, Part C: Journal of Mechanical Engineering Science, 232(21): 3954-3963 (2018).

[5] Wang Y Z, Ren S Y, Li Y. Design and manufacturing of a novel high contact ratio internal gear with a circular arc contact path. International Journal of Mechanical Sciences, 153-154: 143-153 (2019).

[6] Chen Y Z. Line Gear. Beijing (China): Science Press, 2014. (in Chinese)

[7] Chen Y Z, Xing G Q, Peng X F. The space curve mesh equation and its kinematics experiment. 12th IFToMM World Congress, Besançon, France, 2007.

[8] Chen Z, Zeng M, Fuentes-Aznar A. Computerized design, simulation of meshing and stress analysis of pure rolling cylindrical helical gear drives with variable helix angle. Mechanism and Machine Theory 153: 1-19 (2020).

[9] Litvin F L, Tsay C B. Helical gears with circular arc teeth: simulation of conditions of meshing and bearing contact. Transactions of the ASME 107: 556-564 (1985).

Bolt Load Calculation for Hybrid Joint Considering Friction and Nonlinear Stiffness under Combined Thermal and Mechanical Loads

Tiren He [1, a, *], Jifeng Xu [1,b], Kaige Zhu [1,c] and Haijun Wang [1,d]

[1.] Beijing Aircraft Technology Research Institute, Beijing Key Laboratory of Civil Aircraft Structures and Composite Materials, Beijing, 102211 P.R. China

Email: [a,*]hetiren@comac.cc, [b]xujifeng@comac.cc, [c]zhukaige@comac.cc, [d]wanghaijun@comac.cc

Abstract. An improved analytical method for calculating the bolt load distribution in single-lap or double-lap multi-bolt composite-metal joint structure considering material nonlinear behaviour and thermal stress is proposed in this paper. The analytical method is based on the stiffness method and the composite material mechanics. The bolt load ratios calculation results of a three-bolt joint structures by the improved method is compared with the results from finite element (FE) method and the relative error less than 7%. The great agreement verifies the accuracy of the analytical method. Besides, this improved has more considering factors and higher accuracy than traditional stiffness method and less calculating cost than FE method.

Keywords: Composite-metal mechanical joint; Thermal stress; Bolt load calculation;

1. Introduction

In the structural design of aircraft, composite-metal mechanical joint is widely used in the connection structure between outer wing and central wing since its high reliability, load-carrying capacity and convenience to disassemble. However, bolt load distribution and the location of the stress concentration bolt, which are closely related to the failure behaviour, in the composite–metal joint structure are difficult to assign especially under combined thermal-mechanical loads. There are mainly three ways to calculate the load distribution of joints structure: analytical method[1], finite element (FE) method[2] and experiment tests[3]. FE method can simulate nearly all effects in the joint structure with more computational costs[4]. Experimental tests has poor economy and difficult to accurately measure the bolt load[5]. The analytical method has few calculation costs but the calculation accuracy is unsatisfactory owing to less factors considering[6]. The analytical method based on the spring method is extended to investigate the effects of clearance [7] and hole location errors [8]. An improved stiffness method considering the thermal stress, friction effects and nonlinear behaviour is developed in this paper for accuracy calculation of bolt load in the joint structures.

2. The improved stiffness method

A single-lap five-bolt joint connection structure shows in Fig.1 is discussed to investigate the improved analytical method of bolt load distribution based on stiffness method. The numbers of bolts in the drawing from left to right are 1#-5#. The upper plate A has the same material as all the titanium alloy bolts, and the lower connecting plate B is made of IM7/8552te laminate.

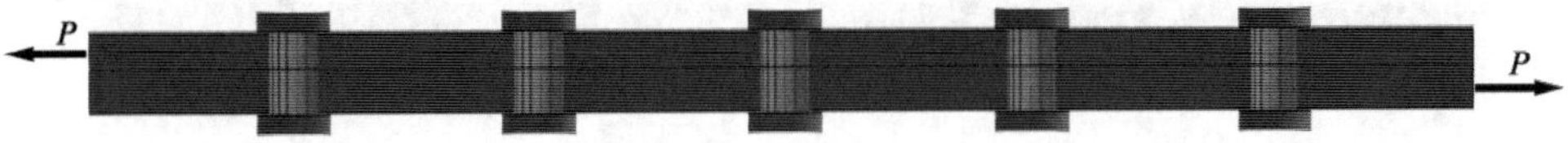

Figure 1. The schematic diagram of single-lap five-bolt joint connection structure.

Friction force. In the initial stage of loading, the static friction between the plates is the dominant factor of load rise. According to the classical Coulomb law, the plates begin to slide relative to each other only when the external load is greater than the static friction force F_0, as illustrated in Eq. 1:

$$P \geq \mu \cdot F_p = F_0, \tag{1}$$

where $\boldsymbol{P}$ is external load, μ is the static friction coefficient between laminates, F_p is the tightening force, F_0 is the static friction force. After the connection plates begins to slide relative, the deformation of connected plate caused by friction can be calculated by sliding friction force and flexibility:

$$u^f = C^f \cdot F^f, \tag{2}$$

where u^f is the deformation of connected plate around the bolt, C^f is the shear flexibility of connected plate around the bolt, F^f is the friction force around the bolt. C^f can be calculated as Eq. 3:

$$C^f = t/\left(A^b \cdot G_{XZ}\right), \tag{3}$$

where t is thickness of laminate plate, A^b is the contact area between the bolt and laminate plate, which is nearly same as the contact area between gasket or nut (without gasket) and laminate,G_{XZ} is out-plan shear module.

Thus, according to the equilibrium condition of force, there is a general equilibrium equation between bolt load, friction force and external load, as shown in Eq. 4:

$$F_1 + F_2 + F_3 + F_1^f + F_2^f + F_3^f = P, \tag{4}$$

where, F_n is the bolt load of bolt NO.n, F_n^f is the friction force around bolt NO.n.

The load transferred between two adjacent nails of plate A is as follows:

$$\begin{cases} F_{12}^A = P - F_1 - F_1^f \\ F_{23}^A = P - F_1 - F_2 - F_1^f - F_2^f \\ \vdots \\ F_{n-1,n}^A = P - (F_1 + \cdots + F_{n-1}) - \left(F_1^f + \cdots + F_{n-1}^f\right) \end{cases} . \tag{5}$$

The load transferred between two adjacent nails of plate B is as follows:

$$\begin{cases} F_{12}^B = F_1 + F_1^f \\ F_{23}^B = (F_1 + F_2) + \left(F_1^f + F_2^f\right) \\ \vdots \\ F_{n-1,n}^B = (F_1 + \cdots + F_{n-1}) + \left(F_1^f + \cdots + F_{n-1}^f\right) \end{cases} . \tag{6}$$

According to Eq. 1 to Eq. 6, the load of each bolt, in the upper and lower connecting plates between adjacent bolts and the external load on the whole connecting structure are shown in Fig. 2.

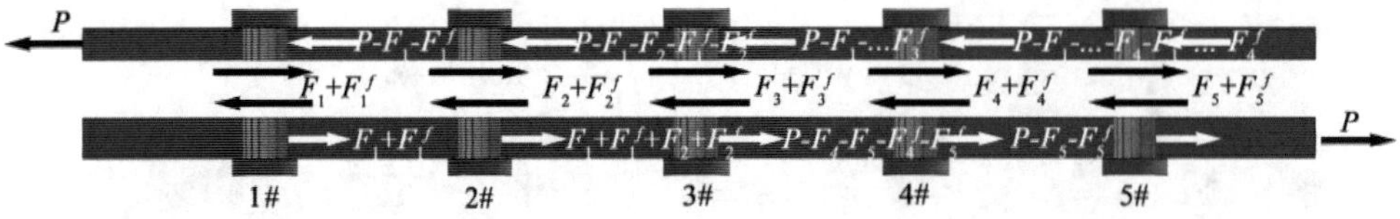

Figure 2. Schematic diagram of loads on the bolt and connecting plates.

Thermal environment. When the temperature changes, the materials with different properties in the composite-metal mechanical joint structure will produce different thermal deformation.

The distance ion both plate A and plate B between two adjacent nails is equal:

$$L_A = L_B. \tag{7}$$

Assuming the deformation of plate *A* is $\Delta^A_{n-1,n}$,the deformation of plate *B* is $\Delta^B_{n-1.n}$,shear deformation of the bolt NO.*n*-1 is Δ_{n-1},shear deformation of the bolt NO.*n* is Δ_n.The deformation compatibility equation is shown in Eq. 8:

$$L_A + \Delta^A_{n-1,n} + \Delta_n = L_B + \Delta^B_{n-1,n} + \Delta_{n-1}. \tag{8}$$

By substituting Eq. 7 into Eq. 8, the compatibility conditions of multi row single row nails can be obtained as follows:

$$\Delta^A_{n-1,n} + \Delta_n = \Delta^B_{n-1,n} + \Delta_{n-1}. \tag{9}$$

The deformation of bolt and laminate between two adjacent bolts under load is shown in Eq. 10:

$$\begin{cases} \Delta_n = F_n \cdot C^b_n \\ \Delta^A_{n-1,n} = F^A_{n-1,n} \cdot C^A_{n-1,n} + \alpha_A \cdot \Delta T \cdot L_A - C^f_n \cdot F^f_n \\ \Delta^B_{n-1,n} = F^B_{n-1,n} \cdot C^B_{n-1,n} + \alpha_B \cdot \Delta T \cdot L_B - C^f_{n-1} \cdot F^f_{n-1} \end{cases}, \tag{10}$$

where, F_n is the bolt load of bolt NO.*n*; C^b_n is the shear flexibility of bolt NO. *n*; Δ_n is the displacement of bolt NO.*n*; $F^A_{n-1,n}$ and $F^B_{n-1,n}$ are the axial load between bolts NO.*n* and NO.*n*-1 in plates A and B, respectively. $C^A_{n-1,n}$ and $C^B_{n-1,n}$ are axial tension flexibility of plates A and B between bolts NO.*n* and NO.*n*-1, respectively. $\Delta^A_{n-1,n}$ and $\Delta^B_{n-1,n}$ are deformation of plates A and B between bolts NO.*n* and NO.*n*-1, respectively. α_A and α_B are coefficient of thermal expansion of plates A and B, respectively. The deformation diagram between two adjacent bolts is shown in Fig. 3.

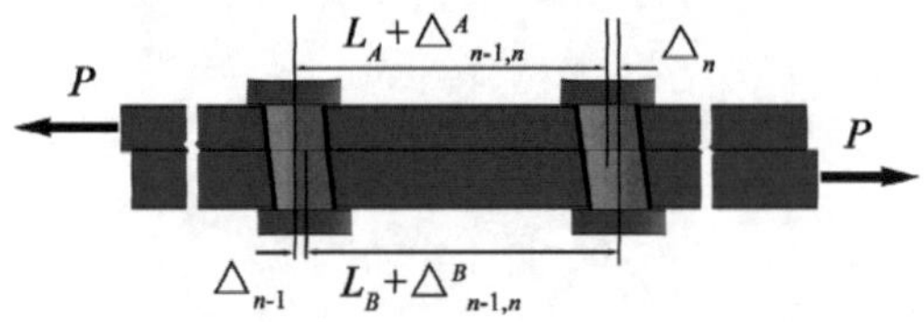

Figure 3. The deformation diagram between two adjacent bolts

Nonlinear deformation of single bolt. The nonlinear deformation properties of single bolt was captured by a tension test of single bolt joint specimen. The stress and strain can be calculate as in Eq. 11:

$$\sigma = P/(D \cdot t)\,, \quad \varepsilon = a \cdot U/D. \tag{11}$$

where σ and ε are stress and strain of the joint structure, *U* is the displacement deformation of the joint area, *D* is the diameter of bolt, *t* is the thickness of laminate plate, *a* is a constant depends on the connection form, a=1 for single-lap, a=2 for double-lap. There are four stages in the deformation process as shown in Fig. 4.

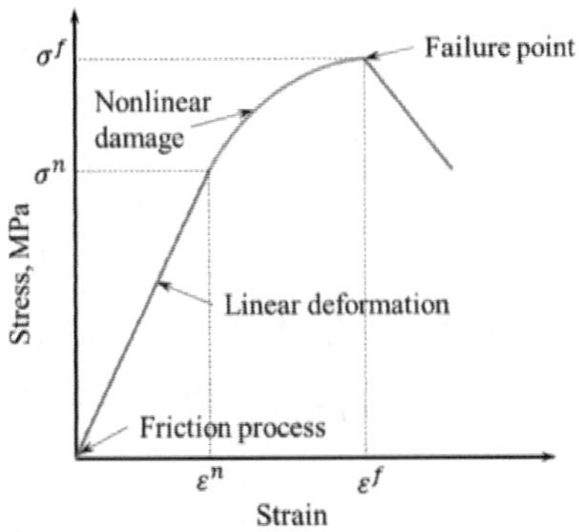

Figure 4. The four stages of stress-strain curve in single bolt test

The connection flexibility of bolt in linear deformation was calculated as shown in Eq. 12:

$$C^b = \frac{\delta}{P} = \frac{(t_A+t_B)^2}{E_b D^3} + 3.72\left(\frac{1}{t_A E_x^A} + \frac{1}{t_B E_x^B}\right), \tag{12}$$

where C_b is the flexibility of bolt, δ is shear deformation, t_A and t_B are thickness of plates A and B, respectively, D is bolt diameter, E_b is young's module, 3.72 is the coefficient about the bolt. The Ramberg-Osgood equation was used to express the nonlinear damage process before the failure point:

$$\varepsilon = \frac{\sigma}{E} + \left(\frac{\sigma}{K}\right)^{\frac{1}{n}}, \tag{13}$$

where E is equivalent module in the linear deformation process, K is secant module, n is nonlinear parameter, K and n were obtained by least squares regression method, the equivalent flexibility can be deduced as shown in Eq. 14.

$$C_n = C^b + \frac{1}{n \cdot D \cdot K} \cdot \left(\frac{P}{D \cdot t \cdot K}\right)^{\left(\frac{1}{n}-1\right)} = C^b + f(K,n). \tag{14}$$

The equivalent flexibility equals to the sum of the flexibility in linear deformation and the nonlinear damage term, which is the function $f(K,n)$ in Eq. 14.

In order to simulate the nonlinear behavior of composite laminate, a progressive loading was adopted in the calculation. The incremental relation is calculated as Eq. 15:

$$\Delta\varepsilon^m = C_n{}^m \Delta N^m, \tag{15}$$

where $\Delta\varepsilon^m$ is incremental strain of composite laminate in mth increment, ΔN^m is the external load. $C_n{}^m$ is the compliance matrix, which deduced through the classical laminated plate theory of composite mechanics.

The flexibility of plates A and B between two adjacent bolts is shown in Eq. 16:

$$C_{n-1,n}^A = \frac{L_{n-1,n}^A}{E_x^A W_{n-1,n}^A t_{n-1,n}^A}, C_{n-1,n}^B = \frac{L_{n-1,n}^B}{E_x^B W_{n-1,n}^B t_{n-1,n}^B}, \tag{16}$$

where E_x^A and E_x^B are tension modules of plates A and B, respectively, $W_{n-1,n}^A$ and $W_{n-1,n}^B$ are the weight of plates A and B, $t_{n-1,n}^A$ and $t_{n-1,n}^B$ are thickness of plates A and B, $L_{n-1,n}^A$ and $L_{n-1,n}^B$ are bolt pitch between bolts NO.n and NO.n-1.

Based on the above equilibrium conditions, compatibility conditions and load displacement relationship, a set of equations with bolt load as unknown parameters can be formed, as shown in Eq. 17:

$$\begin{cases} F_1 C_1^b - F_2 C_2^b = (P - F_1) C_{12}^A - F_1 C_{12}^B + \alpha_A \cdot \Delta T \cdot L_A - \alpha_B \cdot \Delta T \cdot L_B \\ \vdots \\ F_{n-1} C_{n-1}^b - F_n C_n^b = \left[P - \left(\sum_{j=1}^{n-1} F_j\right)\right] C_{n-1,n}^A - \left(\sum_{j=1}^{n-1} F_j\right) C_{n-1,n}^B + \alpha_A \cdot \Delta T \cdot L_A - \alpha_B \cdot \Delta T \cdot L_B \\ F_n = P - \left(\sum_{j=1}^{n-1} F_j\right) \end{cases} . \quad (17)$$

Eq. 18 can be simplified as matrix form in Eq. 17:

$$\boldsymbol{CF}=\boldsymbol{D}, \quad (18)$$

where $\boldsymbol{C}$ is the matrix related to the flexibility of connecting plates A and B and bolts, $\boldsymbol{F}$ is the vector composed of unknown bolt loads, and $\boldsymbol{D}$ is the deformation of connection plates A and B due to external load. Then the matrix form of the final calculation equations of pin load is as follows:

$$\boldsymbol{F} = \boldsymbol{C}^{-1}\boldsymbol{D}. \quad (19)$$

3. Finite element method

A FE model of the joint structure in Fig.1 is established in ABAQUS with 15544 elements. Fig. 5 shows the bolt connection area. For the purpose of simulating the nonlinear deformation of fastener, the plastic constitutive of titanium alloy is defined in the FEM model and geometric nonlinearly is considering in the model. The FE model is meshed with first-order continuum rectangular shape element (C3D8I). The surface-to-surface contact constraints is set in the contact area between the bolt and the connecting plate and the plate between the upper and lower connecting plates. A totally external load of 100KN applied on the right end surface of the upper plate which is the composite laminate and a fixed boundary condition is defined on the left end.

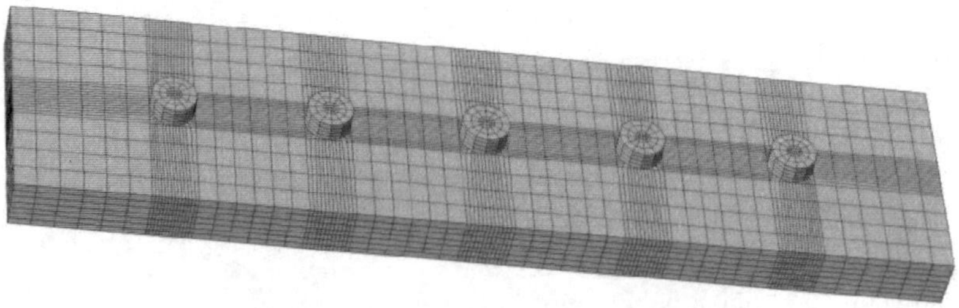

Figure 5. The FE model of five-bolt-joint

4. Results and discussion

Fig. 6 comparing the bolt load distribution ratios calculated by the improved analytical method and FE method. The largest relative errors between analytical method and FE method at #1 bolt is 6.96% because all factors were considered in the FE model, such as the contact between the bolt and the plate and the 3D effect of the joint members. As the error is acceptable, the calculation result is dependable and the improved analytical method is reliable.

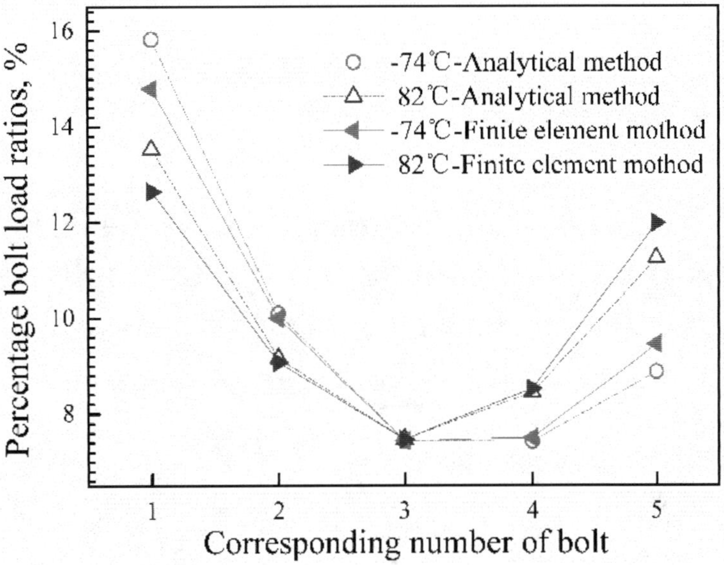

Figure 6. Comparison of bolt load ratios of different method at various temperature

Acknowledgments
This work was sponsored by the Project funded by China Postdoctoral Science Foundation under Grant NO. (2019M651629) and National Natural Science Foundation of China under Grant NO. (11702328). Such support is gratefully acknowledged.

References

[1] Xiang J W, Zhao S W, Li D C, et al. An improved spring method for calculating the load distribution in multi-bolt composite joints J. Composites Part B: Engineering. 117(2017)1-8.

[2] Chen K K, Liu L Q, Wang H. Analysis of Through-the-Thickness Stress Distribution in Thick Laminate Multi-Bolt Joints Using Global-Local Method J. Journal of Shanghai Jiaotong University. 18(2013) 326-333.

[3] Yang K C, Hsu R J, Hsu C F. Effect of End Distance and Bolt Number on Bearing Strength of Bolted Connections at Elevated Temperature J. International Journal of Steel Structures, 13(2013) 635-644.

[4] Lyu Y F, Li G Q, Wang Y B, et al. Bearing behavior of multi-bolt high strength steel connections J. Engineering Structures. 212(2020) 110510.

[5] Zhang D C, Wang G, H F H, et al. Load-transferring mechanism and calculation theory along engaged threads of high-strength bolts under axial tension J. Journal of Constructional Steel Research. 172(2020) 106153.

[6] Jadhav S M, Gadade A M. Micro-Mechanical Based Load Distribution in Multi Bolt Single Lap Composite Joint J. Materials today: proceedings. 16(2019) 668-676.

[7] Liu F R, Shan M J, Zhao L B, et al. Probabilistic bolt load distribution analysis of composite single-lap multi-bolt joints considering random bolt-hole clearances and tightening torques J. Composites Structures. 194(2018) 12-20.

[8] Lecomte J, Bois C, Wargnier H, et al. An analytical model for the prediction of load distribution in multi-bolt composite joints including hole-location errors J. Composite Structures. 117(2014) 354-361.

IOP Conf. Series: Materials Science and Engineering 1102 (2021) 012016 doi:10.1088/1757-899X/1102/1/012016

Influencing Factors of the Ultimate Tensile Bearing Capacity of the Composite Laminated Square Tube Single-bolt Joint

Xiujie Zhu[1,a], Jian Zheng[1,b], Chao Xiong[1,c,*], Junhui Yin[1,d] and Qiongjie Tang[2,e]

[1] Army Engineering University, Shijiazhuang campus, Shijiazhuang, China
[2] 75222 troops of PLA, China

[a]sduzxj@163.com, [b]zhengjian@163.com, [c,*]xiongchao@tsinghua.org.cn, [d]yuanzhidao@163.com,[e]201835684@mail.sdu.edu.cn

Abstract. The three-dimensional progressive damage finite element model was established to accurately simulate the failure process and modes of the composite laminated square tube single-bolt joints. The effects of layup angle, layup ratio and bolt diameter on the ultimate bearing capacity of joints were studied using the validated numerical model, and the reinforced flange gasket was designed according to the damage characteristics of the laminated square tube wall. The results showed that the ultimate bearing capacity of the composite laminated square tube bolt connection structure was relevant to the proportion of 0 degree layers, which increased with the increase of the bolt diameter. The ultimate bearing capacity were increased by the flange gasket reinforcement about 15%.

Keywords: composite laminated square tube; single-bolt joint; tensile experiment; numerical simulation

1.Introduction

The composite laminated square tubes with compression molding has good overall performance and high geometric accuracy, which are often used as the main load-bearing member in the box beam [1-3], the wind turbine blade [4], the launch vehicle instrument compartment [5], helicopter wings and aircraft rotors [6] When the composite laminated square tube transmits a large concentrated load or bears an impact load, connecting with each other through bolt joints is commonly application.

In this paper, the tensile bearing capacity experiments of the composite laminated square tube single-bolt were carried out, and the three-dimensional numerical model based on energy evolution was established in ABAQUS/Explicit. The influence of the layup angle and bolt diameter on the ultimate bearing capacity were analyzed, and the reinforced flange gasket was designed to improve the bearing condition. The references for engineering application were provided.

2.Experiment and simulation

Specimens and experimental method. The definition of the geometric parameters of the composite laminated square tube specimen was shown in Figure 1. The following of the design principle of geometric parameters of mechanical connection structure of composite laminate ($e \geq 3d$, $B \geq 4d$)[7] could ensure that the specimen does not suffer catastrophic failure such as tube wall shear and splitting, which determined the geometric parameters of the composite laminated square tube single-bolt connector: e=35 mm, B=50 mm.

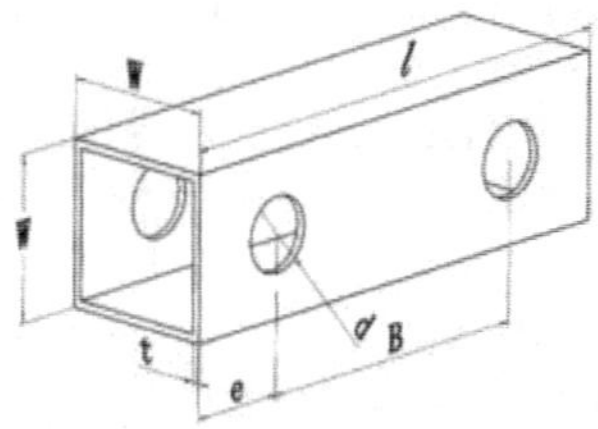

Figure 1. Definition of geometric parameters of specimens

The layup of the composite laminated square tubes for experiment was [(0/45/90±)$_3$/±45], outermost layer of carbon fiber orthogonal twill weave cloth. The total length of the formed tube fittings was 780 mm, and the dimensional accuracy can be controlled within ±0.03 mm. The specimen was drilled by engraving machine. The diameter of the opening hole was the same as the diameter of the bolt, and the hole wall had no visible delamination phenomenon, and the quality was better, as shown in Figure 2.

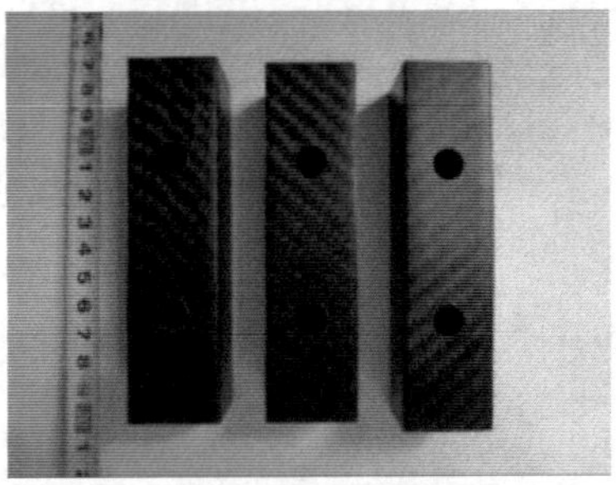

Figure 2. Three specimens for experimenting

The machined metal fixture was used to apply a tensile load, as shown in Figure 3(a). The experiment was carried out on INSTRON5982 universal experimenting machine, and the loading rate was all 2 mm/min, as shown in Figure 3(b). The material of metal fixture and bolt fastener were 35 CrMo alloy steel.

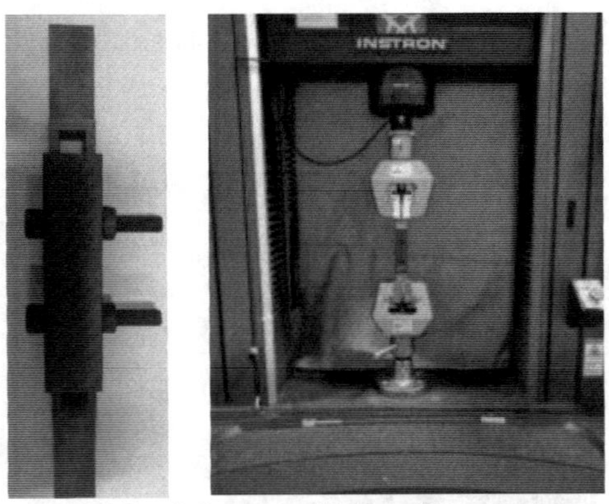

(a) Metal fixture (b) Loading mode

Figure 3. Experimental set of composite laminated square tube bolt connector

Finite element model. The finite element model of overall structure was established in ABAQUS/Explicit, and the symmetrical boundary condition was applied, as shown in Figure 4. The single layer thickness of composite material was 0.15 mm, the outermost carbon fiber twill fabric was replaced by a 0°layer and a 90° layer, and the thickness was 0.125 mm. Sixteen SC8R continuous shell elements were arranged along the wall thickness direction, and the sequence of laying from the inside to the outside of the tube wall was [±45/0/90]$_4$. The Hashin criterion was applied for determining the initiation of damage, and four failure modes of fiber stretching, fiber compression, matrix stretching and matrix compression were distinguished. An energy-based degradation model of material properties was selected. The 0.01 mm thick viscous interface element was inserted between COH3D8 two adjacent layers. the secondary stress criterion was used as the interlayer damage initiation criterion.

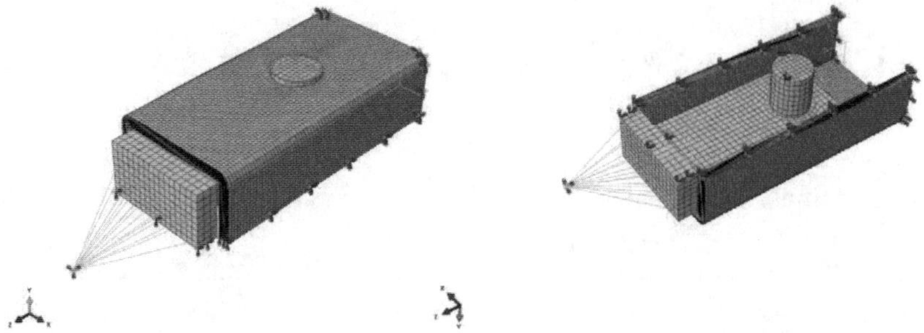

Figure 4. Finite element model of composite laminated square tube bolt connection structure

Validation of simulation. Figure 5 shows the load-displacement curves obtained by the experiment and finite element method. It can be found that the variation trend and peak value of load-displacement curve obtained by experiment and simulation are similar. The ultimate load and connection stiffness obtained by the experiment were slightly lower than the results of numerical simulation due to the material inhomogeneity of the CFRP square tube, the assembly gap and the slip of the fixture during loading. Figure 6 showed the process of tensile failure of specimens. The stress concentration on the loaded side of the bolt hole was severe, and the composite material underwent uplift and delamination deformation, which was basically consistent with the experimental phenomenon.

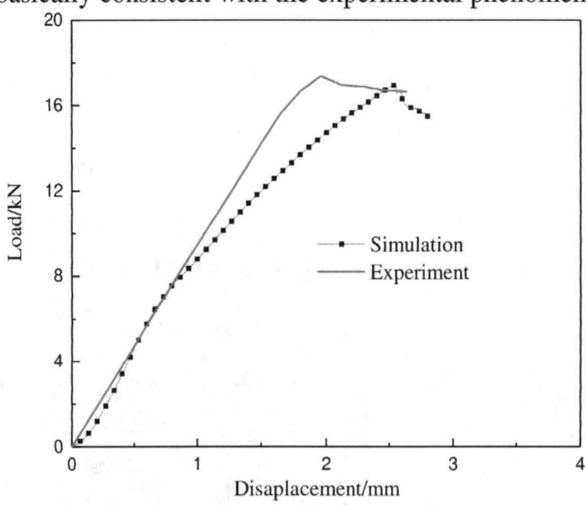

Figure 5. Experimental and simulating load-displacements

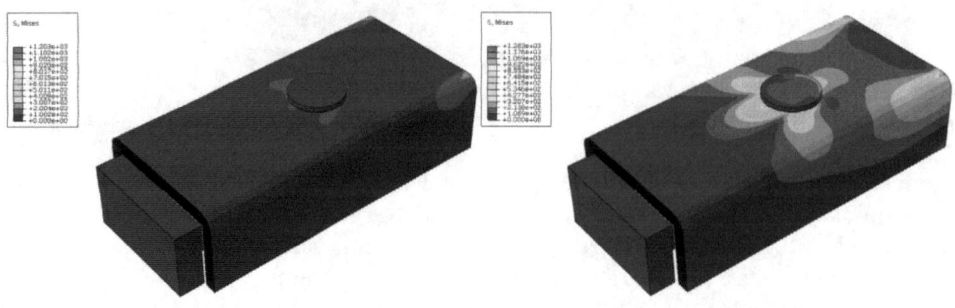

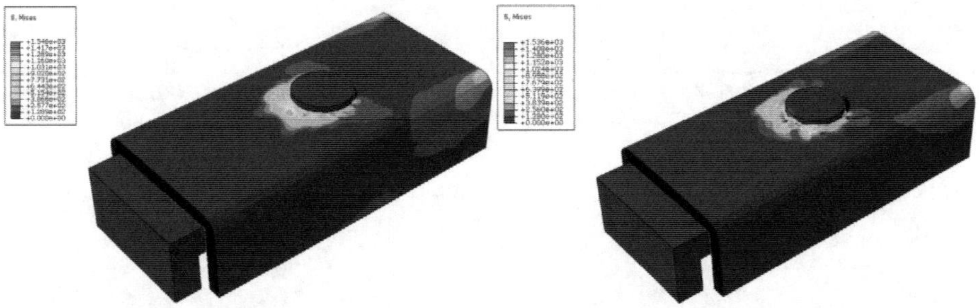

Figure 6. Failure process of composite laminated square tube single-bolt joint

3.Influencing factors of the ultimate tensile bearing capacity of the composite laminated square tube single-bolt joint

Effect of layups. For further analysis of the influence of the angle and proportion of layups on ultimate bearing capacity, the layups [±θ]4, [0/±θ/90]sand [θ/0/-θ]]s were selected. According to numerical model in section 2.2, the finite element model of the corresponding composite laminated square tube bolt connection structure was established, and the tensile ultimate bearing capacity was calculated, and the variation curves of the tensile ultimate bearing capacity with the ply angle of the composite laminated tube with different layers were obtained, as shown in the Figure 7. It could be found that in the condition that the ply angle θ was equal, the ultimate bearing capacity of laminated composite laminated square tube bolt joint with $[0/\theta/0/\text{-}\theta]_4$ layup was the largest.

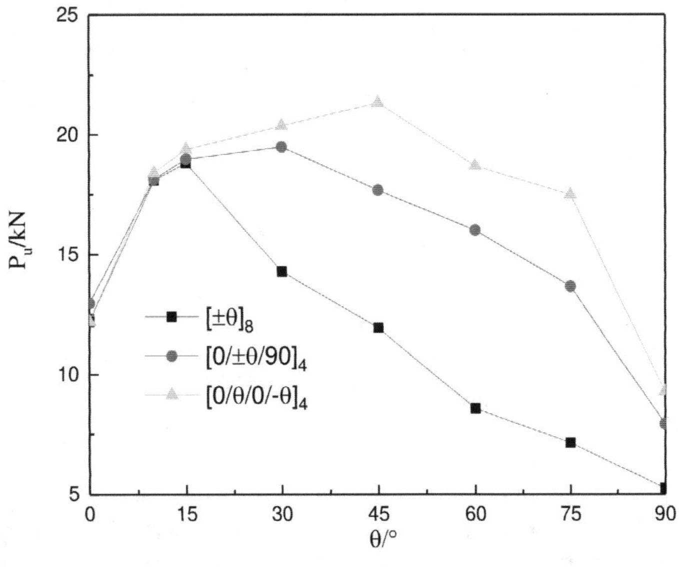

Figure 7. The ultimate bearing capacity composite laminated square tube change with the layups

Effect of bolt diameter. The finite element model of composite laminated square tube mechanical connectors with open hole diameter d of 4 mm, 6mm, 8mm, 10mm, 12mm and 14 mm were established. in each model, the composite laminated square tube is laid in the same way as in 2.2, and the geometric dimensions are shown in Table 1.

Table 1 Geometric parameters of specimens

Specimens	*e/mm*	*B/mm*	*d/mm*	*l/mm*	*W/mm*	*t/mm*
D4	20	30	4	70		
D6	25	30	6	80		
D8	30	40	8	100	30	2.5
D10	35	50	10	120		
D12	45	60	12	150		
D14	50	70	14	170		

$e \geq 3d$, $e \geq 4d$

The following formula was used to calculate the local extrusion failure strength of the joints:

$$\sigma_c = \frac{P_u}{dt}. \tag{1}$$

P_u was the ultimate bearing capacity, σ_c was the failure strength, d was the bolt diameter, t was the thickness of the wall of laminated tube. The variation of σ_c change with d was shown in Figure 8. As the bolt diameter increased, the ultimate bearing capacity of the composite laminated tube bolt connection structure increased nonlinearly. P_u fitted as a power function about the d with a power exponent less than 1. The local extrusion failure strength of the joints decreased with the increase of bolt diameter.

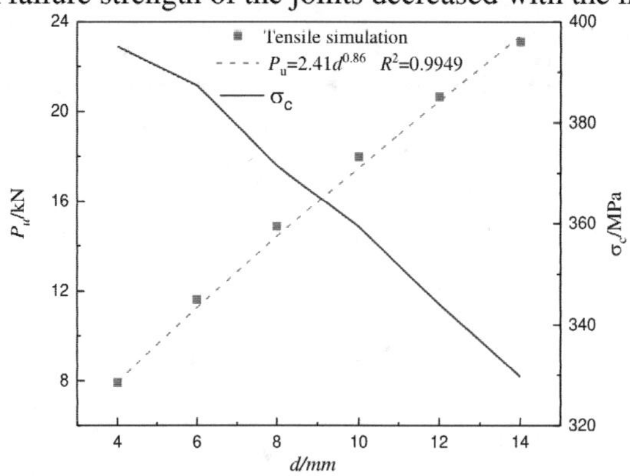

Figure 8. Variation of ultimate bearing capacity and local compressive failure strength with bolt diameter

Effect of reinforced flange gasket. From the results of the gradual damage analysis of the finite element model in section 2.3, it can be seen that the bending deformation of the bolt causes the stress concentration of the inner wall of the tube, resulting in the damage extending from the inner side to the outer side of the tube wall. Based on this phenomenon, considering the structural characteristics of composite laminated square tube and the assembly process of single bolt connection structure, the bolt connection structure of composite laminated square tube established in 2.2 was selected as the reinforcement object. In the case of keeping the diameter of the opening hole unchanged, a flange gasket was designed and installed on the inner side of the tube wall, as shown in Figure 9. The thickness of the flange and the protrusion were both 0.5mm, the flange diameter was 11 mm, the height of the protrusion was 2.5 mm, and the 9 mm diameter bolt was used to realize the gapless assembly with the flange gasket. Through the reinforcement of flange gasket, the stress concentration on the inner side of tube wall could be reduced, and the ability of hole edge to resist delamination failure could be improved.

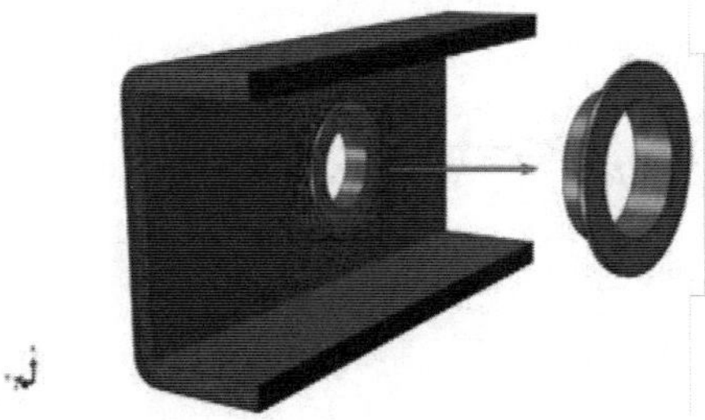

Figure 9. Flange Gasket Enhancement Scheme

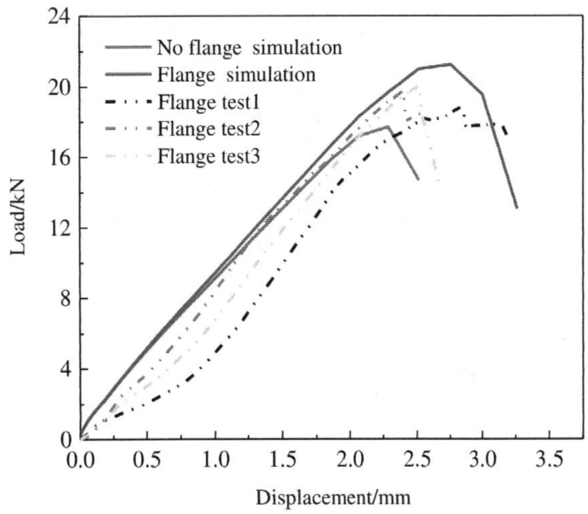

Figure 10. Load-displacement curves of flange gasket reinforced connectors experimented and simulated

The finite element model of composite laminated square tube established in section 2.2 was selected. After adding flange gasket, the progressive damage analysis was carried out to extract the load-displacement curve, which was compared with the connectors without adding flange gasket, as shown in Figure 10. It can be found that when the diameter of the opening hole is constant, the ultimate bearing capacity of the flange gasket reinforcement scheme reaches 21.24 kN, and increases by 21%. Table 2 was the experimental results of the joints adding the flange gasket, the average value was 19.56kN, indicating the design of flange gasket was feasible.

Table 2. The ultimate bearing capacity of reinforced joint of composite laminated square tube with single bolt and flange

Tset	Ultimate load/kN	Average/kN	Dispersion/%
1	18.80		
2	19.79	19.56	2.8
3	20.08		

4.Conclusion

Due to the effect of the bending deformation of the bolt, the stress concentration of the inner wall at the opening of the composite laminated square tube was more serious, and the damage began to expand from the inner side of the tube wall to the outer side of the tube wall. According to this phenomenon, the flange gasket reinforcement scheme was designed. The ultimate bearing capacity of the joints under extrusion load was relevant to the proportion of 0°layers, but the tensile splitting failure would occur under the extrusion load at the opening of all composite laminated tubes with 0° laying, so the 0°laying layer should be dispersed along the wall thickness direction.

References

[1] Tong Ruijin. Research on the numerical method of the composite cross-section thin-walled rod with closed cross-section, Chongqing: Chongqing Jiaotong University, 2009.

[2] Thuc Phuong Vo, Jaehong Lee. Flexural-torsional behavior of thin-walled composite box beams using shear-deformable beam theory, Engineering Structures, 2008, 30: 1958-1968.

[3] Zhanming Qin, Liviu Librescu. On a shear-deformable theory of anisotropic thin-walled beams: further contribution and validations, Composite Structures, 2002, 56: 345-358.

[4] Bao Nengsheng, Cao Renjing, Ye Zhiquan. Finite element analysis of vibration characteristics of wind turbine blade structure, Journal of Solar Energy, 2000, 21 (1): 77-81.
[5] Deng Zhongmin. Design foundation of aircraft composite materials, Beijing: Beijing University of Aeronautics and Astronautics Press, 2015: 2-6.
[6] JEGLEY D C, WU K C,PHELPS J E, et al. Structural efficiency of composite struts for aerospace applications, Journal of Spacecraft and Rockets, 2012, 49(5):915-924.
[7] Wang Yaoxian, Mechanics and Structural Design of Composites, Shanghai: East China University of Technology Press ,2014:159.

AUTHOR INDEX